D0821842

Electrochemical Techniques

for

Inorganic Chemists

Electrochemical Techniques

for

Inorganic Chemists

J. B. HEADRIDGE
Department of Chemistry
University of Sheffield

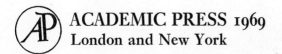
ACADEMIC PRESS 1969
London and New York

ACADEMIC PRESS INC. (LONDON) LTD
BERKELEY SQUARE HOUSE
BERKELEY SQUARE,
LONDON, W1X 6BA

U. S. Edition published by
ACADEMIC PRESS INC.
111 FIFTH AVENUE,
NEW YORK, NEW YORK 10003

Library of Congress Catalog Card Number: 78-85461
SBN 12–335650–4

PRINTED IN GREAT BRITAIN BY
ABERDEEN UNIVERSITY PRESS

Preface

Although the most electropositive metals have been prepared for decades by electrolytic reduction from their molten salts and isolated papers have appeared on the preparation of salts containing elements in less-common higher and lower oxidation states by electrolytic oxidation or reduction, the inorganic chemist would not generally include electrochemical techniques among those which he would use to study the chemical reactions and structures of his compounds. Yet most inorganic reactions involve either complex formation or electron transfer and much information on these phenomena can be readily obtained through the use of electrochemical techniques such as polarography and voltammetry.

Analytical chemists have been using electrochemical techniques for over fifty years, but it is only in the last few years that certain inorganic chemists have realized how useful these techniques can be for investigating redox reactions. In this monograph I show how direct potentiometry, polarography and voltammetry, cyclic voltammetry, chronopotentiometry and controlled potential electrolysis and coulometry can be usefully employed by the inorganic chemist. For certain classes of compounds, electrochemical measurements should now be made along with infrared, e.s.r., n.m.r. and mass spectrometric measurements as a matter of course.

I have deliberately refrained from including much detail on the construction and circuitry of electrochemical instruments, for many excellent textbooks on electro-analytical chemistry which contain full descriptions of equipment are available. I can particularly recommend Lingane, "Electroanalytical Chemistry", second edition, and Meites, "Polarographic Techniques", second edition. Also I do not develop from first principles most of the equations which appear in this book as their derivation is adequately covered in electrochemical texts already published. The emphasis here is on the application of these techniques for the study of inorganic substances in aqueous solutions, in non-aqueous solvents and in molten salts.

Sheffield
June, 1969

J. B. HEADRIDGE

v

Contents

Contents

CHAPTER 1

Introduction

THE inorganic chemist is interested in knowing the formal electrode potential associated with a particular oxidation or reduction, and the number of electrons involved in the electron transfer reaction. From a theoretical standpoint, a correlation between formal electrode potentials, and substituent constants, orbital energies and bonding properties for a series of related compounds is of particular interest. For example, such studies have been made by Shriver and co-workers on substituted borazines,[1] by Shriver and Posner on iron cyanide —BX_3 adducts,[2] by Olson and co-workers on transition metal-dithiolene complexes,[3] and by Hall and Russell on ferrocene derivatives.[4]

The preparative inorganic chemist is interested in knowing whether a particular compound, often newly prepared, can be readily oxidized or reduced to a related compound. When the formal electrode potential for a particular couple is known, it is then possible to select a suitable oxidizing agent to convert the reduced form of that couple to the oxidized form, or a suitable reducing agent to convert the oxidized form of that couple to the reduced form. If a species X can be reduced successively to X^- and X^{2-} then from a knowledge of the formal electrode potentials for the couples X/X^- and X^-/X^{2-}, a suitable reducing agent can be selected to bring about the reduction of X to X^-, while another more powerful reducing agent can be chosen to bring about the reduction of X to X^{2-}. Similar information can be obtained for oxidations. The oxidizing or reducing agents are selected from a table of formal electrode potentials for many couples, obtained in the same solvent containing the same concentrations of other salts and neutral molecules in addition to the species constituting the couple. This is best illustrated by the following.

In anhydrous acetonitrile, which is 0·1 M in tetraethylammonium perchlorate, the formal electrode potentials of the couples samarium (III)/samarium (II), as perchlorates, and samarium (II)/samarium amalgam are approximately −0·98 V and −1·69 V, respectively, *versus* the aqueous saturated calomel electrode (aq. S.C.E.). In 0·1 M tetraethylammonium perchlorate in acetonitrile the formal electrode potentials for the couples aluminium (III)/aluminium amalgam and potassium (I)/potassium amalgam are approximately −1·4 V and −1·96 V *versus* the aq. S.C.E.[5] Samarium (II) in acetonitrile could then be prepared by shaking samarium (III) perchlorate in acetonitrile with aluminium amalgam, while shaking samarium (III) perchlorate in acetonitrile with potassium amalgam would reduce it to samarium metal dissolved in mercury. Controlled potential electrolysis of samarium (III) at -1.3 V *versus* the aq. S.C.E. would also produce samarium (II) quantitatively, and at -2.0 V *versus* the aq. S.C.E., samarium amalgam.

Cyclic voltammetry of *bis*-π-(3)-1,2-dicarbollylnickel (IV), $Ni(B_9C_2H_{11})_2$, in acetonitrile, 0.1 M in tetraethylammonium prechlorate, reveals two reversible one-electron reduction peaks with formal electrode potentials of $+0.25$ and -0.59 V *versus* the aq. S.C.E.[6] These correspond to the reduction of the nickel (IV) complex to the nickel (III) and nickel (II) complexes, respectively. One equivalent of cadmium metal quantitatively reduces the nickel (IV) complex in acetone to the nickel (III) complex; this would be expected, since the values of the formal electrode potentials in acetone will be similar to those given above. In acetone, the formal electrode potential of the cadmium (II)/cadmium amalgam couple is -0.18 V *versus* the aq. S.C.E.[5] and correcting this to solid cadmium gives a value of about -0.25 V *versus* the aq. S.C.E.

A more powerful reducing agent was required for the production of the *bis*-π-(3)-1,2-dicarbollynickel (II) anion and hydrazine in aqueous alkaline solution was employed. The standard electrode potential of the hydrazine/ nitrogen couple shows that hydrazine in alkaline solution is suitable.[7]

$$N_2H_4 + 4OH^- \rightleftharpoons N_2 + 4H_2O + 4e$$

$E^0 = -1.40$ V *versus* aq. S.C.E. Its usefulness, of course, depends on there being a suitable kinetic pathway for equilibrium to be established.

Standard and Formal Electrode Potentials

All chemists are familiar with standard electrode potentials but the meaning of formal electrode potential is now explained.

If a rod of metal, M, is immersed in an aqueous solution of metallic ions, M^{n+}, of unit activity, and if the potential of this electrode is measured *versus* the normal hydrogen electrode (zero volts by definition), then the standard electrode potential, E^0, for the couple M^{n+}/M is this measured potential.

If the activity of the metallic ions is other than unity, then the electrode potential is given by the Nernst equation

$$E = E^0 + \frac{RT}{nF} \ln \{M^{n+}\} \tag{1.1}$$

where $\{M^{n+}\}$ is the activity of the ion, M^{n+}.

If the couple consists of two ions, $M^{(a+n)+}$ and M^{a+}, of equal activity in the presence of an inert electrode such as platinum, then the electrode potential measured *versus* the N.H.E. is the standard electrode potential for the couple $M^{(a+n)+}/M^{a+}$. If the activities are not equal then

$$E = E^0 + \frac{RT}{nF} \ln \frac{\{M^{(a+n)+}\}}{\{M^{a+}\}}. \tag{1.2}$$

If the electrode reaction involves hydrogen ions, it may be represented, for example, by

$$MO_x^{(a-2x+n)+} + 2xH^+ + ne \rightleftharpoons M^{a+} + xH_2O.$$

Then

$$E = E^0 + \frac{RT}{nF} \ln \frac{\{MO_x^{(a-2x+n)+}\}\{H^+\}^{2x}}{\{M^{a+}\}} \tag{1.3}$$

and the electrode potential is the standard electrode potential when all the ions are of unit activity.

From the practical point of view, the use of standard electrode potentials is restricted. While an electrode potential can be readily measured it cannot often be converted to a standard electrode potential, because in contrast to the formal concentrations information on the activities may be completely lacking. It is then necessary to determine or calculate formal electrode potentials.

The formal concentration or formality, F, is the number of formula weights of the substance per litre [8]; it expresses a concentration in terms of the formula weights of the substance *actually added* to the solution per litre of the solution. No information is being given about the concentrations of the molecular or ionic species *actually present* in the solution. For example, the substance added to a solution might be complexed in the solution with the result that the actual molarity of the substance which remains uncomplexed would be lower than its formality and might not readily be determined.

If a rod of metal, M, is immersed in an aqueous solution of unit formality in the ions M^{n+} of ionic strength I, then the potential of this electrode is the formal electrode potential $E^{0\prime}$ of the couple $M(n)/M$. If the concentration of M^{n+} is other than $1F$, the electrode potential, E, is given by the expression

$$E = E^{0\prime} + \frac{RT}{nF} \ln M(n) \tag{1.4}$$

where $M(n)$ is the formal concentration of M^{n+}, provided that the ionic strength remains constant at I. If the solution contains a reagent, C, which complexes M^{n+}, then the formal concentration of C should be constant and at least a hundred times that of M^{n+} if the above equation is to hold.

Modified Nernst equations are also readily written for couples in which two ions, or two ions and the hydrogen (or hydroxyl ion), are involved. Activities are simply replaced by formal concentrations. However, the ionic strength of the solutions should remain constant, and the concentration of any complexing agent in the solutions should be constant and at least a hundred times that of the metal ion(s) involved in the couple.

When quoting formal electrode potentials, in contrast to standard potentials, the concentrations of other ions and neutral species other than those constituting the couple must be given. To illustrate this point the standard electrode potential and some formal electrode potentials for the iron (III)/iron (II) couple are given in Table 1.

TABLE 1

Standard or Formal Electrode Potentials for the Iron (III)/Iron (II) Couple [9]

		Solution composition			
	0	1 M HClO$_4$	1 M HCl	1 M H$_3$PO$_4$	1 M HF
Potential (V *versus* N.H.E.)	+0·771	+0·73	+0·70	+0·44	+0·32

The value in the first column is the standard electrode potential. With E^o the activities of the hydrated ferrous and ferric ions must be equal. In 1 M hydrofluoric acid, however, the activity of the hydrated ferric ion is very much less than the activity of the hydrated ferrous ion, although the formal concentration of iron (III) is equal to that of iron (II). This arises because iron (III), but not iron (II), is strongly complexed by fluoride ion.

Formal electrode potentials are frequently obtained in solutions where the concentrations of the soluble species constituting the couple are about $10^{-3}F$. The standard or formal electrode potentials of some familiar couples in aqueous solution are shown in the Appendix, but a more detailed list is given in Reference 10.

It will have been noted that much of the preceding discussion has been orientated towards aqueous solutions, which is merely because much less information is available about formal electrode potentials in non-aqueous solvents. However, formal electrode potentials for couples in several non-aqueous solvents are given in Chapter 8. The factors, which influence the values of the formal or standard electrode potentials of metallic complexes have been reviewed by Perrin[11] and by Buckingham and Sargeson.[250]

Formal Electrode Potentials and *n* values

For many compounds information on formal electrode potentials and *n* values can be obtained quickly, and most of the inorganic chemists who have used electrochemical techniques during the last ten years have used them only for this purpose. The author feels that most inorganic chemists will only be interested in using electrochemical techniques if they can get useful information in a few hours, and he has concentrated on the means of getting this information and on the types of compounds for which this redox information is easily obtained.

When both the oxidized and reduced forms of the couple are available its formal electrode potential can often be readily measured by a direct potentiometric procedure, where the couple constitutes one of the electrodes in a cell and the other electrode is a reference electrode. This method is described in Chapter 2. When only one form of the couple is available, as is very often the case, polarography, voltammetry, cyclic voltammetry or chronopotentiometry can, in many circumstances, be employed to measure formal electrode potentials, and these techniques are described in Chapters 3 and 5. The n value is the number of electrons involved in the oxidation or reduction, and some information on the n values can be obtained from polarograms, voltammograms and cyclic voltammograms; this also will be discussed further in Chapters 3 and 5. However, controlled potential coulometry is a more satisfactory way to determine n values and this is described in Chapter 6. Thin-layer electrolysis as a means of determining n values is a very recent development and is mentioned in Chapter 5.

Other Information from Electrochemical Investigations

Electrochemical techniques can also be employed to obtain much useful information in addition to formal electrode potentials and n values. If an electron transfer process in polarography is preceded or followed by a relatively slow chemical reaction, an analysis of the polarographic wave obtained under different conditions of concentration of the depolarizer, temperature, and so on, will yield much information about the reaction mechanism and the structure of intermediate ions and molecules. This is a fascinating field of study which has been ably reviewed by Vlček.[12, 13] With time and patience, chronopotentiograms,[14] cyclic voltammograms [15] and controlled potential coulometric data [16] can also be analysed to give similar detailed information. However, these detailed studies will not be discussed at length in this monograph since the inorganic chemist who wishes to obtain quick redox information about his compounds will not usually want to investigate the polarograms, chronopotentiograms, and so on, at such depth.

Although a complete investigation of an electron transfer process coupled with one or more chemical reactions using the techniques just mentioned may take weeks, much useful information about electrochemical reactions coupled with chemical reactions can be obtained more quickly in a partial study by combining polarography and voltammetry, cyclic voltammetry, chronopotentiometry and controlled potential electrolysis and coulometry with other techniques such as ultraviolet and infrared spectroscopy, e.s.r. spectroscopy and gas chromatography.

As yet this type of approach has only been adopted by a few workers, such as Dessy *et al.* (see Chapter 8 [211-217]) and Smith *et al.* (see Chapter 5 [82, 83]), but as electrochemical techniques become more widely used by inorganic chemists, more studies of this type can be expected.

Electrochemical Techniques and Analytical Chemistry

Although inorganic chemists are only now beginning to use electrochemical techniques, in particular polarography and voltammetry, to an appreciable extent, these techniques have been widely used by analytical chemists over the last twenty years. Several books have been written on polarography [17-20] from the analytical standpoint where the emphasis has been on quantitative analysis for the height of a polarographic wave is usually directly proportional to the concentration of the electroactive species producing the wave. Amperometric titrations are based on polarography and voltammetry, while in potentiometric titrations use is made of electrode potential measurements to determine equivalence points. Controlled potential electrolysis is also employed as a quantitative technique, where the *n* value is known and the concentration of the electroactive substance is determined using Faraday's laws of electrolysis. The analytical applications of these techniques are described by Kolthoff and Lingane,[17] Milner,[18] Meites,[19] Crow and Westwood,[20] Lingane [21] and Rechnitz,[22] and in reference 23.

Further Remarks

Processes for the electro-winning and electro-refining of metals such as aluminium, and the industrial electrochemical synthesis of substances such as chlorine, sodium chlorate and sodium hydroxide will not be discussed in this book, since many of these processes are already familiar to the inorganic chemist. For a recent appraisal of these processes in the United States, an article by Wenglowski [24] should be consulted.

Much information on formal electrode potentials in aqueous solution is available from books and papers on electro-analytical chemistry and this information is easily found, because the electrochemical techniques which have been used are mentioned in the titles. However, other electrochemical information of interest to the inorganic chemist is frequently hidden away within inorganic papers and no clue to the existence of this material is given in the titles. Most of the information collected by the author from such inorganic papers has come from metal-hydrocarbon and sulphur-ligand chemistry.

CHAPTER 2

Direct Potentiometry

THIS method for determining electrode potentials will already be familiar to most inorganic chemists. It is of more limited use for determining formal electrode potentials than polarography, voltammetry, cyclic voltammetry and chronopotentiometry because, as stated in Chapter 1, both species constituting the couple are required in known formal concentrations. However, direct potentiometry has been and still is being widely used by analytical and physical chemists to determine equilibrium constants through the measurement of the electrode potentials of appropriate couples usually in aqueous solutions. The measurement of electrode potentials and the use of the Nernst equation allows the activities of particular ions to be calculated and later used in appropriate equations for determining stability constants and solubility products. These aspects of direct potentiometry are discussed by Rossotti and Rosotti [25]; many stability constants appearing in the Chemical Society's publication [10] are determined in this way.

Inorganic chemists working with compounds which they have synthesized may wish to use direct potentiometry occasionally, and therefore the technique is described briefly in this chapter.

Apparatus

The requirements are a suitable H-cell containing a reference electrode, an electrode constituted from the couple under examination, and a potentiometer. These are shown in Fig. 1.

The operation of a potentiometer is familiar and will not be described here; full instructions are given in manufacturers' literature.

The H-cell shown in Fig. 1 is suitable for aqueous solutions. The salt bridge is a gel which is usually made by heating 3 g of Bacto-agar in 100 ml of saturated (25°C) potassium chloride solution on a steam bath until the Bacto-agar has dissolved. The mobile solution is poured into the horizontal cross-arm of the cell, which has been previously heated in an oven to about 80°C. The cell is left clamped at room temperature with the cross-arm in a vertical position until the gel has set. If the introduction of chloride ions into the solution compartment will cause difficulties at a later stage, 0·1 M sodium perchlorate solution, rather than saturated potassium chloride solution, may be used to make the gel.

The reference electrode is an aqueous saturated calomel electrode and is prepared by pouring a few millilitres of pure mercury into the bottom of the reference electrode compartment, followed by a paste of calomel intimately mixed with potassium chloride crystals in a saturated solution of potassium chloride. The compartment is then filled with saturated potassium chloride solution until the gel is completely covered, and the compartment securely stoppered. A length of glass tubing, closed at the lower end and with a

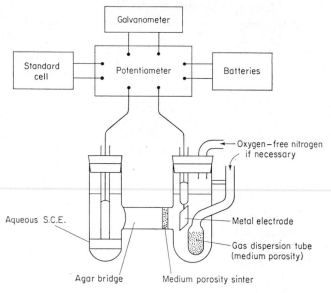

FIG. 1. Equipment for determining electrode potentials in aqueous solutions using direct potentiometry.

piece of platinum wire sealed through it, is half filled with mercury and pushed through a hole in the stopper until the platinum wire is completely immersed in the mercury at the bottom of the compartment. Connection to the reference electrode is then made very simply by immersing a wire from the potentiometer in the mercury of the glass tubing. When the cell is not in use, the end of the gel next to the solution compartment must be covered with the appropriate salt solution to prevent drying-out of the gel. The cell should be kept immersed in a water-bath at constant temperature, usually 25°C.

If the electrode potential of a couple consisting of a metal in the presence of its ions is to be determined, then that part of the working electrode which is immersed in the aqueous solution is made from the required metal or is a platinum electrode with the metal plated on to it. When both the oxidized and reduced species are in solution, the working electrode is usually made of platinum. As with the reference electrode, for ease of connection to the potentiometer, it is often convenient to seal a platinum wire, spot-welded to a platinum foil electrode, through the lower end of a length of glass tubing, and half fill the glass tubing with mercury.

The gas dispersion tube in the solution compartment is used when oxygen has to be removed from the solution. This is often necessary, since the reduced form of many couples will react with oxygen. The passage of a stream of oxygen-free nitrogen through the solution for a few minutes will remove the dissolved oxygen and provision should then be made to pass the nitrogen over the surface of the solution to prevent re-entry of oxygen. Commercial "oxygen-free" nitrogen usually contains traces of oxygen, which should be removed by passage through chromous chloride solution or over hot copper.[26]

A steady value of the cell potential is often obtained more rapidly, if the platinum electrode is platinized. This is conveniently done in the following way.[27] Electrolyse a solution of chloroplatinic acid 1% aqueous (W/V) containing 8 mg of lead acetate per 100 ml with the electrode as the cathode and a platinum coil anode for five minutes under an applied potential of two volts. The polarity is then reversed every minute for a further five minutes. The electrolyte is replaced with a solution of sulphuric acid (2+98) and electrolysis is continued for a further two minutes with the platinum electrode as anode. Finally the electrode is rinsed with distilled water and roasted in a Bunsen flame until red hot.

When a couple in acidic solution has an electrode potential more negative than zero volts *versus* the N.H.E., mercury should be used instead of platinum as the working electrode. On mercury a large overvoltage (about 1 V) is required for the reduction of hydrogen ion; with platinum, the overvoltage is small.[28] The formal electrode potential of, for example, the chromium (II)/chromium (III) couple in acidic solution can be determined on mercury. On platinum, however, the reaction

$$H^+ + Cr^{2+} \rightarrow \tfrac{1}{2}H_2 + Cr^{3+}$$

proceeds rapidly and a reasonably constant ratio of chomium (II) to chromium (III) cannot be maintained.

When the electrode potential of a couple is determined in a non-aqueous solvent, where traces of water are of no consequence, a cell similar to that shown in Fig. 1 may be used. However, a sodium perchlorate gel should be employed, since potassium chloride is insoluble in many non-aqueous solvents and deposition of potassium chloride crystals in the sinter can lead to an undesirable high cell resistance.

When the electrode potential of a couple has to be determined in a strictly anhydrous solvent, a cell again similar to that in Fig. 1 may be used, provided

that the solvent is used throughout the cell. The problem with this arrange-
ment is finding suitable gelling agents for the cross-arm. Pantony [29] has
stated that fully-methylated methyl celullose (5% W/V) is a suitable gelling
agent for saturated solutions of tetraethylammonium perchlorate in dimethylformamide and dimethylsulphoxide. Methyl cellulose can also be used as a
gelling agent with glacial acetic acid [30] and sulpholane,[31] but is not suitable
for acetonitrile and propylene carbonate.[32] A reference electrode which can
be used in most non-aqueous solvents is silver/silver chloride/saturated silver
chloride+tetraethylammonium chloride solution. Other reference electrodes
for non-aqueous solvents are mentioned by Hills [33] and Takahashi.[34]

Since methyl cellulose is not a universal gelling agent for non-aqueous
solvents, a cell suitable for all anhydrous solvents and containing an aqueous
saturated calomel electrode (saturated sodium chloride) is shown in Fig. 2.

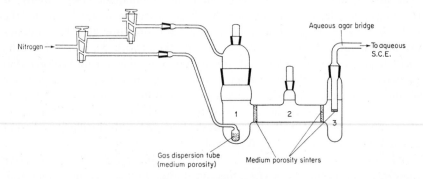

FIG. 2. A suitable cell for determining electrode potentials in anhydrous solvents. When
potential data are to be obtained the stopper at the top of compartment (1) is removed and
the electrode is placed in the solution. The bridge is an aqueous gel of 3·5% (W/V) agar
in 0·1 M sodium perchlorate solution.

The anhydrous solvent containing the species under investigation is placed
in compartments (1), (2) and (3). With this cell any trace of water diffusing
through the small sinter has no possibility of reaching compartment (1)
within a reasonable period of time. Saturated sodium chloride rather than
saturated potassium chloride is used in the reference electrode, since potassium
perchlorate will precipitate at the interface with the agar, 0·1 M sodium
perchlorate gel, if saturated potassium chloride is employed. The potential
of the saturated calomel electrode (sodium chloride) at 25°C is −0·008 V
versus the saturated calomel electrode (potassium chloride). Obviously many
cells of different designs can be made for particular systems but, essentially,
they are all similar to those described here.

If the couple forms a reversible system (see Chapter 4), a steady value for
the cell potential is usually attained quickly in aqueous solution. However, if

the electron transfer processes at the working electrode are slow, this constitutes an irreversible system and many minutes may have to elapse before a steady reading is attained. If the electron transfer reactions are very slow, however, an external e.m.f. set to oppose the e.m.f. of the cell will provide very little current in the galvanometer attached to the potentiometer, even if the working electrode potential is appreciably removed from the equilibrium potential, that is the electrode potential dictated by the ratio of the activities of the oxidized and reduced forms through the Nernst equation. It will be virtually impossible to balance the e.m.f.'s, and such couples are very irreversible.

A reversible couple is one such that, when a potential very slightly different from its equilibrium electrode potential is imposed on the electrode, the ratio of the oxidized and reduced forms quickly take up the new value set by the Nernst equation. Obviously this can only happen when the electron transfer reactions are fast.

Some electron transfer reactions are very slow indeed, for example

$$SO_4^{2-} + 4H^+ + 2e \rightarrow SO_2 + 2H_2O.$$

The standard electrode potential of this couple cannot be measured potentiometrically and it has to be determined from standard free energies obtained by methods other than electrochemical.[35]

Fortunately many couples form systems which are not too irreversible and their electrode potentials can be determined by the potentiometric method. However, Popov [36] has stressed that equilibrium conditions are often established much more slowly in non-aqueous solvents than in water and this fact must be borne in mind when using the direct potentiometric method.

Occasionally it may be possible to determine the formal electrode potential of a couple, when only the oxidized or reduced form is available, by potentiometric titration provided that a suitable titrant is available. If the reduced form of a couple is titrated with an oxidizing agent, then the formal electrode potential of the couple is the potential of the working electrode, when half of the reductant has been titrated. This potential is readily obtained from the potentiometric titration curve. The formal electrode potentials for substituted ferrocene/substituted ferricinium ion couples in aqueous acetic acid have been obtained in this way, the ferrocene derivatives being titrated with potassium dichromate solution.[37]

CHAPTER 3

Polarography and Voltammetry

VOLTAMMETRY is a special type of electrolysis, where one of the electrodes in the voltammetric cell is a reference electrode of appreciable surface area and the other is a working electrode of very small area. This working electrode can be made of any conducting material but is usually a small cylinder or disc of carbon, gold or platinum, or a dropping mercury electrode. When the working electrode is a dropping mercury electrode, the technique is called polarography. The reference electrode has a large surface area compared with the working electrode so that the reference electrode is not readily polarized, that is the concentration of ions at the surface of the electrode should be almost identical to their concentration in the bulk of the solution, when small currents flow through the cell. In this way the potential is constant.

In polarography and voltammetry, information about the solution in which the working electrode is immersed is obtained by studying polarographic and voltammetric waves on a graph of current flowing through the cell against applied potential.† Usually in d.c. polarography and voltammetry an increasing potential is slowly applied to the cell at a rate of 0·05–0·3 V min⁻¹, and the cell current is measured or recorded as the potential increases. The potential scan is usually over a restricted range within the region of +3 to −3 V *versus* the aq. S.C.E. and the currents recorded are seldom in excess of 50 μA.

Polarographs

A simplified diagram of the apparatus required for two-electrode d.c. polarography and voltammetry is shown in Fig. 3, and these instruments are called polarographs although voltammographs would be a more correct name.

In most instruments the potential slide wire R_1 is calibrated by means of a standard cell (circuitry not shown in Fig. 3) and the required potential across

---

† Strictly speaking, voltammetry is the general name for the technique, and it includes polarography. However, in this book the name polarography will be used to describe the technique of obtaining current-voltage recordings where the small electrode is a dropping mercury electrode, while voltammetry will describe the similar technique for all other types of small electrodes.

the cell can be varied by altering the position of contact T. In recording polarographs, T is moved along R_1 by an electric motor; the resulting current is allowed to pass through a resistance R_2 and the potential drop across R_2 is measured with a potentiometer; in recording polarographs, this is a potentiometric recorder. Widely varying currents can be measured by switching different resistances into the position occupied by R_2 on the diagram. Commercial polarographs, of which there is a wide selection, have controls for selecting the initial voltage, the voltage scan and the sensitivity, that is μA per full scale deflection on the recorder.

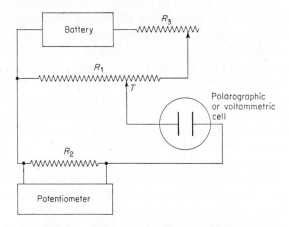

FIG. 3. A simplified diagram of a polarograph.

With a two-electrode d.c. polarograph, the potential of the working electrode with respect to the reference electrode is slightly less than the potential being applied to the cell if the current through the cell is other than zero. This is because there is a small iR drop in the cell and if the cell resistance is less than 500 Ω and the current less than 2 μA, then the iR drop (1 mV) can often be neglected. For aqueous solution the cell resistance is usually less than 500 Ω, but for non-aqueous solvents the cell resistance is frequently between 1,000 and 10,000 Ω and the iR drop is then appreciable. The potential of the working electrode is then obtained by correcting the applied potential for the iR drop across the cell, and an example of this correction is given on page 29. In the most exact work, another correction should be made for the iR drop across the measuring resistance, R_2, although this correction seldom exceeds 2 mV with a 2·5 mV full-scale potentiometric recorder.

The need to correct for iR drop across the cell can be avoided if a controlled-potential d.c. polarograph is used. Three electrodes are then required in the cell, namely, a reference electrode, a working electrode and an| auxiliary electrode. The measured cell current is that passing between the working and auxiliary electrodes, but the potential of the working electrode *versus* the reference electrode at any time during the scan is set by means of the controls on the polarograph. Conditions are arranged so that the potential between the working and auxiliary electrodes constantly changes in such a way as to ensure that no appreciable current flows between the reference and working electrodes. Under these conditions, there is a negligible potential drop in the solution between the working and reference electrodes except in solutions of very high resistance, and the potential of the working electrode *versus* the reference electrode is, at any time, that read from the polarograph.

Commercial instruments have been recently discussed fully in the excellent book by Meites,[19] and the author will not repeat that information here. From personal experience, the author can strongly recommend the Sargent Model XV Recording Polarograph (E.H. Sargent and Company, Chicago, U.S.A.). A new instrument by Beckman (Beckman Instruments Inc., Fullerton, California, U.S.A.) based on an operational amplifier, the Electroscan 30, should be an attractive buy to the inorganic chemist since it can be operated as a controlled-potential polarograph and also for cyclic voltammetry and chronopotentiometry (see Chapter 5) and controlled potential electrolysis as a preparative technique (see Chapter 8).

The author considers that the inorganic chemist who wishes to obtain polarographic data is best served by a conventional, slow scan d.c. polarograph. It must be remembered that polarography is at present primarily an analytical technique, and great efforts have been made to improve upon the sensitivity of the d.c. technique described above. Instruments such as the differential cathode ray polarograph, which is a rapid scan instrument,[38] and the pulse polarograph [39] will allow the quantitative determination of ions at concentrations two orders of magnitude lower than with the usual d.c. polarograph, but are less suited to the inorganic chemist, who is interested in formal electrode potentials and n values.

Cells

The cells required for two-electrode polarography and voltammetry are similar to those used for direct potentiometry. For work in aqueous solutions, a cell like that in Fig. 1 is suitable. In polarography, where it is the intention of the investigator to obtain data at potentials negative to the aq. S.C.E., the solution is de-oxygenated before recording a current–voltage graph, and, during the recording, nitrogen is passed over the surface of the solution to prevent the re-entry of oxygen, which is reduced at potentials negative with respect to the aq. S.C.E. In aqueous solution, oxygen is reduced first to hydrogen peroxide and then to water.

$$O_2 + 2H_2O + 2e \rightarrow H_2O_2 + 2OH^-$$
$$O_2 + 2H_2O + 4e \rightarrow 4OH^-$$

In aprotic solvents, oxygen is reduced first to superoxide ion and then to peroxide ion.[40, 41]

$$O_2 + e \rightarrow O_2^-$$
$$O_2^- + e \rightarrow O_2^{2-}$$

These reduction waves can frequently interfere with the waves obtained from the substance under investigation; hence the necessity for removing oxygen.

With non-aqueous solvents, cells similar to those in Figs 1 and 2 may be used, and the information given in Chapter 2 for the use of these cells with non-aqueous solvents is equally relevant for polarography and voltammetry. For work with anhydrous solvents, the nitrogen gas must be oxygen-free and dry. This is conveniently done by passing oxygen-free nitrogen, first through a column of 4A molecular sieves (Union Carbide) and then through a column of phosphorus pentoxide. If the solvent under investigation is fairly volatile, the nitrogen should be saturated with the vapour of this solvent before it passes into the cell. With aqueous solutions, plastic tubes may be used to bring the nitrogen to the cell, but with many non-aqueous solvents, it is best to use all-glass lead-in tubes (as in Fig. 2) after the presaturating solvent bottle. This is because the vapours of many organic solvents attack plastic tubing and can carry forward oxidizible and reducible impurities into the cell.

The most widely used working electrode is the dropping mercury electrode (polarography). When this electrode cannot be used, because the potential range being studied is more positive than the potential for the oxidation of mercury, rotating platinum, gold, glassy carbon, pyrolytic graphite or carbon paste electrodes may then be employed. Of these, the rotating platinum electrode is the most popular and is readily prepared by sealing a piece of platinum wire, of approximately 0·4 mm diameter, through the lower end of a length of glass tubing, and cutting the wire so that the piece which protrudes through the end is about 3 mm in length. Obviously, platinum wires of different surface areas can be made to suit individual needs. The glass tubing is half-filled with mercury and placed in a chuck, which is driven at a constant speed (often 600 r.p.m.) by a suitable electric motor.

With controlled-potential polarographs, a third electrode, the auxiliary electrode, must be incorporated in the cell. A suitable cell for aqueous solutions and for non-aqueous solvents where a trace of water is of no consequence is shown in Fig. 4. The working electrode is shown here as a dropping mercury electrode but, of course, a solid electrode can be used if necessary. This cell can also be used with a non-aqueous solvent throughout the cell if a suitable gelling agent is available for a non-aqueous salt bridge; under these circumstances a non-aqueous reference electrode replaces the aq. S.C.E. The conical glass projection, P, points towards the working electrode, the open end of P being placed as close as possible to the working electrode; under these conditions the correction to be applied to the potential of the working electrode for the potential drop in the solution between the

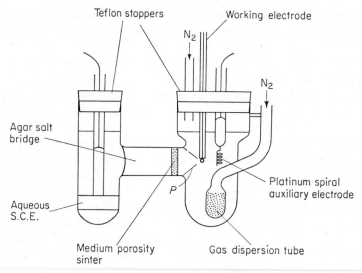

FIG. 4. A suitable cell for controlled-potential polarography with aqueous solutions.

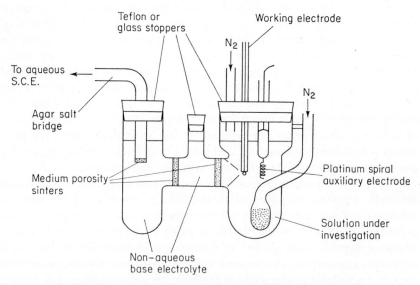

FIG. 5. A suitable cell for controlled-potential polarography with anhydrous solutions and an aqueous saturated calomel electrode as reference electrode.

orifice of P and the working electrode is negligible unless the resistance of the solution is very large. A small hole is made in the upper side of projection P, so that all air can escape from the cone with the ingress of solution, although in less exact work the conical projection P is unnecessary. When the reference electrode is an aqueous saturated calomel electrode and the solution under investigation is an anhydrous solvent, then the cell may be of the type shown in Fig. 5. If further details on cells and electrodes are required, Meites[19] should be consulted.

The Interpretation of Current-voltage Curves

Typical current-voltage curves, that is polarograms or voltammograms, are shown in Figs 6,7,8,9 and 10. In all these figures, the steep rise in current at positive potentials (A) is caused by one of the following: (1) oxidation of the electrode material, (2) oxidation of the solvent or (3) oxidation of the base electrolyte (almost always the anion of the base electrolyte). For example, on a polarogram of 0·1 M sodium perchlorate in dimethylformamide, A results from oxidation of the mercury of the dropping mercury electrode.

$$2Hg \rightarrow Hg_2^{2+} + 2e.$$

On a voltammogram (rotating platinum electrode) of 0·1 M sodium perchlorate in dimethylsulphoxide, A is caused by oxidation of the solvent. On a voltammogram (R.P.E.) of 0·1 M potassium iodide in water, A is due to oxidation of iodide ion.

$$3I^- \rightarrow I_3^- + 2e.$$

In Figs 6–10, the steep rise in current at negative potentials (B) is due to the reduction of either the solvent itself or the base electrolyte (almost always the cation of the base electrolyte). For example, on a polarogram of 0·1 M lithium perchlorate in glacial acetic acid, B results from reduction of the acetic acid.

$$CH_3COOH + e \rightarrow CH_3COO^- + \tfrac{1}{2}H_2.$$

On a polarogram of 0·1 M sodium chloride in water, B results from reduction of sodium ion.

$$Na^+ + Hg + e \rightarrow Na(Hg).$$

In Fig. 6, C is the residual current line. This line rises slowly towards negative potentials. The residual current is made up of a condenser current, which is a consequence of the charging of the electrical double layer at the electrode surface, and of a diffusion current due to traces of reducible or oxidizible substances in the solution.

In Fig. 7, D is a wave due to the reduction of a species deliberately added in low concentration (10^{-5}–10^{-2} M) to the base electrolyte. This species can be a cation, neutral species or anion, for example:

$$Cd^{2+} + Hg + 2e \rightarrow Cd(Hg) \qquad \text{in most solvents}$$

$$O_2 + e \rightarrow O_2^- \qquad \text{in aprotic solvents}$$

$$IO_3^- + 6H^+ + 6e \rightarrow I^- + 3H_2O \qquad \text{in water.}$$

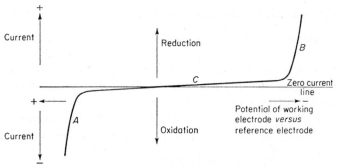

FIG. 6. A current-voltage recording for a base electrolyte.

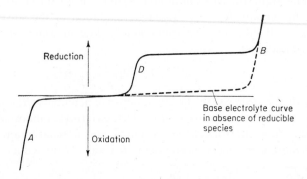

FIG. 7. A current-voltage recording for a base electrolyte plus a reducible species (axes as in Fig. 6).

In Fig. 8, E and F are waves due to the reduction of, say, europium (III) in acetonitrile, first to europium (II) and then to europium amalgam, that is:

$$Eu^{3+} + e \rightarrow Eu^{2+}$$

$$Eu^{2+} + Hg + 2e \rightarrow Eu(Hg)$$

Note that the height of the second wave is twice that of the first, since n, the number of electrons involved in the reduction, is one for the first reduction but two for the second.

In Fig. 9, G is a wave caused by the oxidation of a species added to the base electrolyte in low concentration. For example:

$$(\pi\text{-}C_5H_5)_2Fe \rightarrow (\pi\text{-}C_5H_5)_2Fe^+ + e \quad \text{in most solvents}$$
$$\underset{\text{ferrocene}}{} \qquad \underset{\text{ferricinium ion}}{}$$

In Fig. 10, H is an oxidation wave and J is a reduction wave. A polarogram like this is obtained for tin (II) in 0·1 M aqueous ammonium fluoride solution.

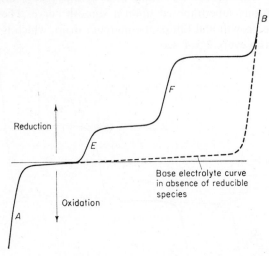

FIG. 8. A current-voltage recording for a base electrolyte plus one species, which is reduced in two steps (axes as in Fig. 6).

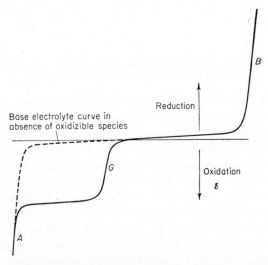

FIG. 9. A current-voltage recording for a base electrolyte plus an oxidizible species (axes as in Fig. 6).

That is:

$$H \text{ results from Sn (II)} \rightarrow \text{Sn (IV)} + 2e$$

$$J \text{ results from Sn (II)} + \text{Hg} + 2e \rightarrow \text{Sn(Hg)}$$

The waves are of equal height since n is two in each case.

It should be appreciated that the current-voltage recordings for solid electrodes are smooth curves, while polarograms show a succession of regular oscillations superimposed upon a smooth curve. These oscillations result from the growth and fall of the mercury drop, which is usually regulated to one drop every 2·5–4 sec.

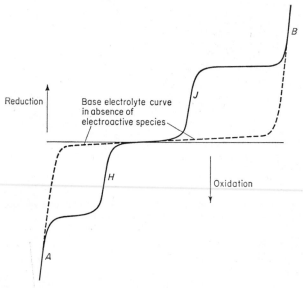

FIG. 10. A current-voltage recording for a base electrolyte plus one species, which can be both oxidized and reduced (axes as in Fig. 6).

Two questions arise from these current-voltage recordings: firstly why are the oxidation and reduction waves S-shaped, and secondly why is a base or supporting electrolyte required? To answer these questions, one can consider, for example, the polarographic reduction of a low concentration (10^{-4} M) of cobalticinium ion in aqueous 0·1 M potassium chloride solution. The formal electrode potential for the cobaltocene/cobalticinium couple is $-0·92$ V versus the N.H.E. or $-1·16$ V versus the aq. S.C.E. At potentials more positive than $-0·9$ V versus the aq. S.C.E., the concentration of cobalticinium ion at the surface of the dropping mercury electrode is 10^{-4} M, whilst the concentration of cobaltocene is virtually zero. However, as the potential of the D.M.E. approaches the formal electrode potential, some of the cobalticinium ions are reduced at the electrode, so as to produce that ratio of cobaltocene to cobalticinium ion dictated for the system by the Nernst equation. Since cobalticinium ion is reduced to cobaltocene

$$(\pi\text{-}C_5H_5)_2Co^+ + e \rightarrow (\pi\text{-}C_5H_5)_2Co$$

a current must flow through the cell. This current increases rapidly as the formal electrode potential is approached. However, the current would soon decrease, if it were not that a steady supply of cobalticinium ions was reaching the electrode surface by diffusion from the bulk of the solution.

At a potential corresponding to the formal electrode potential, the concentrations of both species at the surface of the electrode are equal. At potentials more negative than the formal electrode potential, the ratio of cobaltocene to cobalticinium ions must exceed one according to the Nernst equation, and at potentials more negative than -1.40 V *versus* the S.C.E., the concentration of cobalticinium ions at the surface of the electrode is practically zero.

After the formal electrode potential the current increases more slowly and then reaches a constant value as the potential of the D.M.E. approaches -1.40 V. This constant current is maintained because cobalticinium ions are constantly being brought up to the electrode at a steady rate by diffusion, and the resulting polarographic wave is therefore S-shaped. The current on the plateau of the wave, after correction for the residual current, is the diffusion current for a 10^{-4} M solution of cobalticinium ions in 0.1 M aqueous potassium chloride.

In this argument, it is assumed that the reduction of cobalticinium ions is very rapid. This is indeed the case, and the electrode process is thermodynamically reversible (see Chapter 4).

The base electrolyte (0.1 M potassium chloride) is required in order to keep the cell resistance low, and to suppress completely the electrical migration current which would exist in addition to the diffusion current in the absence of a large concentration of non-reducible cation. The potassium ion does not start to reduce until the potential of the D.M.E. reaches -1.9 V *versus* the aq. S.C.E. The concentration of the base electrolyte should be at least fifty times that of the electroactive substance under investigation.

The equation for a reversible polarographic or voltammetric reduction wave, where the oxidized and reduced species are both in solution, is

$$E_{\text{w.e.}} = E_{\frac{1}{2}} - \frac{2.303RT}{nF} \log \frac{i}{i_d - i} \qquad (3.1)$$

and is obtained by applying diffusion theory to the Nernst equation. $E_{\text{w.e.}} =$ the potential of the working electrode for a current i; $E_{\frac{1}{2}} =$ the half-wave potential, that is the potential corresponding to a position halfway up the wave; n = the number of electrons involved in the reduction; i_d = the diffusion current.

$$2\cdot303\frac{RT}{F} = 0\cdot059 \qquad \text{at } 25°C.$$

With a voltammogram, i_d is replaced by i_l, the limiting current. For an oxidation wave, the equation takes the same form except that the sign after $E_{\frac{1}{2}}$ is now positive. The characteristic S-shaped waves are obtained from a plot of current i against the potential of the working electrode $E_{w.e.}$

Equation (3.1) implies that the shape of a reversible wave is controlled solely by diffusion and the Nernst equation. The electron transfer processes are fast for a thermodynamically reversible system and have no influence on the shape of the wave. However, when the electron transfer processes are slower, the shape of the wave is affected; the wave is less steep, that is more drawn out, and is said to be irreversible. The reason for this is simply that the electro-active species diffusing to the working electrode from the bulk of the solution is not being reduced (or oxidized) fast enough at the electrode surface in the vicinity of the formal electrode potential for the ratio of oxidized to reduced species set by the Nernst equation to be attained. For a reduction wave, however, the rate of reduction does increase exponentially as the potential of the working electrode becomes increasingly more negative after the formal electrode potential (see Chapter 4), and at sufficiently negative potentials the same diffusion current is attained for the irreversible wave as would have been attained sooner if the wave had been reversible.

The reversibility or irreversibility of a wave is of prime importance for the inorganic chemist, because, for a reversible wave,

$$E_{\frac{1}{2}} = E^{0\prime} - \frac{0\cdot059}{n} \log \frac{D_0^{\frac{1}{2}}}{D_R^{\frac{1}{2}}} \qquad \text{at } 25°C \qquad (3.2)$$

where $E_{\frac{1}{2}}$ = half-wave potential; $E^{0\prime}$ = formal electrode potential; n = number of electrons involved in the oxidation or reduction; and D_0 and D_R = the diffusion coefficients for the oxidized and reduced species respectively. D_0 and D_R are a measure of the rate of diffusion of these electro-active species in the solution.

Since D_0 and D_R are not vastly different, and since the ratio of their square roots are within a log term, this log term is zero, or nearly so, and can be neglected. Hence

$$E_{\frac{1}{2}} \simeq E^{0\prime}.$$

The formal electrode potential of a couple is, therefore, obtained from the half-wave potential of a reversible oxidation or reduction wave. However a word of warning about this last statement is appropriate because it is only true if a relatively slow chemical reaction does not precede or follow the rapid electron transfer process.

Consider that a relatively slow chemical reaction precedes the rapid electron transfer process, that is

$$\text{A} \overset{\text{slow}}{\underset{\text{fast}}{\rightleftharpoons}} \text{B} \overset{\text{fast}}{\rightleftharpoons} \text{C},$$

where B, but not A, is electro-active, then the limiting current will be partly or entirely kinetic in character. If the equilibrium concentration of B in the bulk of the solution is negligible, then the current is entirely kinetic in character, but if the equilibrium concentration of B in the bulk of the solution is an appreciable part of the total concentration of A plus B then the limiting current is made up from a diffusion current plus a kinetic current. Under these conditions Eqn (3.2) is invalid.[42]

If it is suspected that the wave is not entirely diffusion controlled, the following test should be applied. For a diffusion wave, the diffusion current is proportional to $h_{corr}^{\frac{1}{2}}$ to a good approximation, where h_{corr} is the head of mercury in the capillary of a dropping mercury electrode, corrected for the back pressure due to interfacial tension at the drop surface.[43] If the limiting current is entirely kinetic in character, then the wave height is independent of h_{corr}. If the limiting current is partly kinetic in character, the wave height will increase as h_{corr} is increased, but not so rapidly as for an entirely diffusion-controlled wave. Therefore, if the current is not proportional to $h_{corr}^{\frac{1}{2}}$, the formal electrode potentials of the couples A/C or B/C must not be equated with the half-wave potential of the reversible wave.

If the rapid electron transfer process is followed by a relatively slow chemical reaction, these can be represented by

$$\text{B} \overset{\text{fast}}{\underset{\text{fast}}{\rightleftharpoons}} \text{C} \overset{\text{slow}}{\rightarrow} \text{D}.$$

The wave will be reversible but again the half-wave potential will be different from the formal electrode potentials of the couples B/C and B/D, for Eqn (3.2) does not hold in this case either. Further details can be obtained from the chemical literature.[44, 45]

An example of this effect is the reduction of B-phenylborazine in 1,2-dimethoxyethane and dimethylformamide.[1] The formal electrode potential can still be calculated from the half-wave potential if the first-order or pseudo first-order rate constant of the following chemical reaction can be determined, but the original paper should be consulted for further details.

A rapid electron transfer process followed by a relatively slow chemical reaction is easily recognized from a cyclic voltammogram (see Chapter 5). These reversible waves, obtained for rapid electron transfer processes preceded or followed by slow chemical reactions, are sometimes called quasireversible waves [45] and they will be described as such in this book.†

† The term a "quasireversible wave" has also been used in the past to describe an irreversible wave which is not totally irreversible (see Chapter 4); this latter use of the word is entirely different from the author's use of the word.

Fortunately a great many inorganic species are oxidized or reduced polarographically or voltammetrically without these complications. For example, complexes whose oxidized and reduced forms are both inert cannot have chemical reactions preceding or following the oxidation or reduction and very often such chemical reactions are ruled out by the chemistry of the systems. However, when a slow chemical reaction could be coupled with the fast electron transfer reaction, a cautious approach should be adopted and the half-wave potential should not be equated with the formal electrode potential without further investigation. If, however, coupled chemical reactions are fast and reversible and the electron transfer reaction is fast, then the half-wave potential of the reversible wave can be equated with the formal electrode potential. The polarographic reduction of, for example, many cadmium complexes to cadmium amalgam is in this category. It is the coupled chemical reactions, whose rates are comparable with the rate of diffusion, which lead to the half-wave potentials being different from the formal electrode potentials, even when the electron transfer reactions are fast. Vlček [12] has shown that quasireversible and other polarographic waves can be satisfactorily analysed to yield much information about the structures of intermediate compounds in the redox and chemical processes associated with these waves, and about the reaction mechanisms, but this work is rather specialized and will not be discussed here.

It will be appreciated from the above, that if a species O is reversibly reduced polarographically or voltammetrically to a species R with no preceding or subsequent slow chemical reactions, then the half-wave potential of the wave is to a good approximation the formal electrode potential for the couple, O/R. If R is now oxidized polarographically or voltammetrically, a reversible wave will result with its half-wave potential again identical or nearly identical to the formal electrode potential. In fact, the same half-wave potential for the reduction wave of O and the oxidation wave of R are conclusive proof that the couple is reversible and that there are no slow chemical reactions preceding or following the electron transfer reaction. If the solution contains both O and R, then a composite wave is obtained, which straddles the residual current line. The composite wave has the same shape as the oxidation and reduction waves, and its half-wave potential is again a good approximation to the formal electrode potential.

If a wave is irreversible, its half-wave potential cannot be equated with the formal electrode potential for the appropriate couple. Even so, some information about the formal electrode potential for the relevant couple associated with an irreversible polarographic wave is obtained from the wave, for the formal electrode potential must be more positive than the half-wave potential of a reduction wave, and more negative than the half-wave potential of an oxidation wave. The reduced form of a reversible couple, whose formal electrode potential is more negative than the half-wave potential of an irreversible reduction wave, will reduce the species under study. The oxidized form of a reversible couple, whose formal electrode potential is more positive than the half-wave potential of an irreversible oxidation wave, will oxidize the species responsible for the wave.

Tests for reversibility

From Eqn (3.1) for a reversible reduction wave, a plot of $-E_{w.e.}$ against $\log(i/i_d-i)$ will produce a straight line of slope, $59/n$ mV at 25°C; for an oxidation wave a plot of $E_{w.e.}$ against $\log(i/i_d-i)$ produces a straight line of slope $59/n$ mV. If the wave is irreversible, and therefore more drawn out, the slope is larger than $59/n$ mV.

Another simpler method, although less exact, since it involves measurements on two points instead of, say, ten for the first method, is to determine the $E_{\frac{1}{4}}-E_{\frac{3}{4}}$ value for the wave. For a reversible reduction wave, $E_{\frac{1}{4}}-E_{\frac{3}{4}} = 56/n$ mV at 25°C, and for a reversible oxidation wave $-56/n$ mV. For an irreversible wave, $E_{\frac{1}{4}}-E_{\frac{3}{4}}$ is, neglecting the sign, greater than $56/n$ mV.

Cyclic voltammetry (see Chapter 5) is a convenient way of testing for the complete reversibility of a couple when only the oxidized or reduced form is available.

The reduction of species to insoluble products

Most of the polarographic and voltammetric waves of interest to the inorganic chemist are for substances whose oxidation or reduction products remain in solution, but occasionally an inorganic species is reduced to an insoluble element. This obviously happens when a metallic ion is reduced to the metal on a solid electrode but some elements, such as iron, are also insoluble in mercury.

Equation (3.1) cannot be used because the reduced species is not in solution, and the relevant equation here for a reversible polarographic reduction wave at 25°C is

$$E_{w.e.} = E_{\frac{1}{2}} - \frac{0 \cdot 059}{n} \log \frac{i_d}{2(i_d-i)}. \tag{3.3}$$

For a voltammogram, i_d is replaced by i_l, the limiting current. The shape of a polarogram for this type of system is shown in Fig. 11. For a reversible wave of this type $E_{\frac{1}{4}}-E_{\frac{3}{4}} = 28/n$ mV; for an irreversible wave $E_{\frac{1}{4}}-E_{\frac{3}{4}}$ will be greater than $28/n$ mV. For a reversible reduction, the half-wave potential is related to the formal electrode potential by Eqn (3.4),

$$E_{\frac{1}{2}} = E^{0\prime} + \frac{0 \cdot 059}{n} \log \frac{C}{2} \tag{3.4}$$

where C is the concentration of the reducible ion in the solution.

2

In theory, it should be possible to determine the formal electrode potential of the appropriate couple from this equation, but when this has been attempted polarographically for ions of chromium, iron, cobalt and nickel in aqueous solution—the metals are known to be insoluble in mercury—no conclusion could be drawn about the formal electrode potentials of the appropriate couples, since the waves were all irreversible.[19] In molten salts, voltammetric reductions to the metals are usually reversible and formal electrode potentials can be obtained using Eqn (3.4) suitably amended for the different temperature.

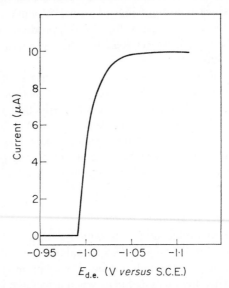

FIG. 11. Calculated polarographic wave for the reversible two-electron reduction of a metal ion to a metal completely insoluble in mercury. The half-wave potential was assumed to be -1.000 V and the diffusion current $10.00\,\mu$A. (Reprinted with permission from Meites.[19])

Determination of n values

Analytical chemists are primarily interested in the magnitude of the diffusion or limiting current, since i_d or i_l is proportional to the concentration of the electroactive species, when the head of mercury for the D.M.E. or the speed of rotation of a solid micro-electrode is kept constant. For polarography, this relationship can be expressed as $i_d = kc$, where k, the proportionality constant, is $708\,n\,D^{\frac{1}{2}}m^{\frac{2}{3}}t^{\frac{1}{6}}$ to a good approximation. i_d = maximum diffusion current expressed in μA; c = concentration in mmoles l^{-1}; n = number of electrons involved in the oxidation or reduction; D = diffusion coefficient of the electroactive species in cm^2 sec^{-1}; m = flow rate of mercury from the capillary in mg sec^{-1}; and t = drop time in seconds.

The maximum diffusion current is the diffusion current measured at the top of the oscillations on a polarogram, that is it is the diffusion current just before the drop falls.

The usefulness of the Ilkovic equation

$$i_d \simeq 708 \, n \, D^{\frac{1}{2}} m^{\frac{2}{3}} t^{\frac{1}{6}} c \qquad (3.5)$$

to the inorganic chemist is that it often enables n values to be determined, when there is doubt about these values. For example, suppose that one has obtained a polarographic reduction wave for a 10^{-3} M solution of a nickel complex, NiX_2^{2-}, where X^{2-} is an organic complexing agent. The $E_{\frac{1}{4}} - E_{\frac{3}{4}}$ value for the wave is 56 mV, which immediately suggests that one is dealing with a reversible or quasireversible wave, where $n = 1$, although it could just be an irreversible two-electron reduction wave. However, it is known that the similar copper complex, CuX_2^{2-}, is reduced to the species CuX_2^{3-}, n being unity for this reduction. To prove that $n = 1$ also for the reduction of NiX_2^{2-}, a polarogram is recorded for a 10^{-3} M solution of CuX_2^{2-} in the same base electrolyte, at the same temperature and with an identical head of mercury as was used for the nickel complex.

For both waves, c, m and t are constant and since the sizes of the two complexes must be about the same $D^{\frac{1}{2}}$ can then be taken as constant; hence $i_d = k'n$ to a good approximation, k' being a constant. If the waves are of approximately equal height, then $n = 1$ for the nickel complex also. When a comparison of wave heights cannot be made because a suitable complex, for which the n value is known, is not available, then controlled potential coulometry should be employed to determine the n value for the new compound.

If two polarographic waves appear on the same polarogram for a single substance, whose electron transfer reactions are free from accompanying slow chemical reactions, then the n values are scarcely ever in doubt, even if both waves are irreversible, as the chemistry of the system can usually only be explained by one particular solution of the polarographic data. For such waves $i_d(1)/i_d(2) = n(1)/n(2)$.

A similar method for comparing wave heights to gain information on n values can be used for voltammetry with rotating solid electrodes. Such voltammograms show limiting currents rather than diffusion currents, since the electroactive species are brought to the electrode by convection as well as by diffusion.

The procedure for intercepting a polarographic or voltammetric wave is therefore as follows.

(1) Is it an oxidation or reduction?
(2) Determine the $E_{\frac{1}{4}}$ value and correct it for iR drop in the cell and across the measuring resistance if necessary.

(3) Determine the corrected $E_{\frac{1}{4}} - E_{\frac{3}{4}}$ value or the slope of a plot of $\pm$ corrected $E_{\text{w.e.}}$ against $\log [i/(i_d - i)]$.

(4) Guess at a realistic value for n from the expected chemistry of the system. Assuming that this n value is correct, is the wave reversible?

Remember that, if the wave is not reversible, you cannot equate the formal electrode potential with the half-wave potential. If the wave is reversible but the rapid electron transfer reaction may possibly be preceded or followed

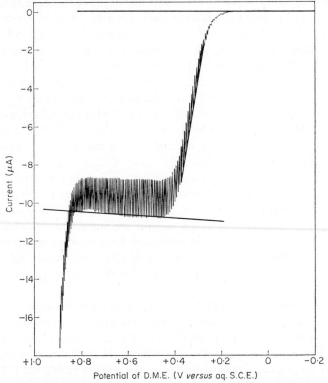

FIG. 12. The polarographic wave for the oxidation of approximately 10^{-3} M ferrocene in nitromethane, which was 0·1 M in tetraethylammonium perchlorate.

by a relatively slow chemical reaction, do not equate the formal electrode potential with the half-wave potential without further investigation using, for example, cyclic voltammetry. A convenient way to measure the half-wave potential from a polarogram is shown in Fig. 12.

The line along the maxima of the oscillations on the plateau of the wave is extrapolated in the direction of the wave. The line along the maxima of the oscillations on the base electrolyte recording before the wave is extrapolated towards the wave; a line is drawn along the maxima of the oscillations in the region of the half-wave potential. A ruler is now pushed along the paper parallel to the current axis until the distance from the first to the third line equals the distance from the third to the second line. The point where the ruler intersects the second line is the half-way potential. The $E_{\frac{1}{4}}$ and $E_{\frac{3}{4}}$ values are obtained in a similar way.

Whether the maximum current or the average current is used in making these extrapolations to obtain the half-wave potential, and so on, is of little consequence provided that the recorder pen has a reasonably fast response to changes in current. If the recorder pen is heavily damped, the mid-points of the oscillations are still the average current but the tops of the oscillations will no longer correspond to the maximum current. The predicted average current is actually six-sevenths of the maximum current, and using average current the constant in the Ilkovic equation is 607 rather than 708.

A voltammogram obtained with a rotating, solid electrode should, of course, be a smooth curve, but minute vibrations of the rotating electrode can lead to a curve showing slight "noise", and the best straight line through this "noise" should be drawn. The cell resistance is conveniently measured with a conductivity bridge (1000 c/sec alternating current). In polarography, the cell resistance for maximum current is determined.

When the procedure for interpreting polarographic and voltammetric data is applied to the wave in Fig. 12 the following results are obtained.

(1) The current is negative. Hence we are dealing with an oxidation.

(2) The uncorrected $E_{\frac{1}{2}}$ value is $+0.324$ V *versus* the aq. S.C.E. The cell resistance is 3000 Ω and the current at the half-wave potential is 5.4 μA, hence the potential drop across the cell is 16.2 mV. The potential drop across the measuring resistance is 0.54 mV. (A full-scale deflection of 250 mm corresponds to 2.5 mV on this recorder. At the half-wave potential the pen has moved 54 mm; hence the potential drop across the measuring resistance in the recorder is 0.54 mV). The corrected $E_{\frac{1}{2}}$ value is therefore $+0.307$ V *versus* the aq. S.C.E. Note that the wave must always become steeper if corrected working electrode potentials are plotted. Considering this, it is easy to see whether to add or subtract the iR drops to obtain the corrected $E_{\frac{1}{2}}$ value.

(3) The corrected $E_{\frac{1}{4}} - E_{\frac{3}{4}}$ value is 60 mV; the slope of the plot of corrected $E_{\text{w.e.}}$ *versus* $\log [i/(i_d - i)]$ is 60 mV. (Errors in measuring $E_{\text{w.e.}}$ values and particularly $E_{\frac{1}{4}}$ and $E_{\frac{3}{4}}$ values for a reversible wave can cause the $E_{\frac{1}{4}} - E_{\frac{3}{4}}$ value or the slope to be a few millivolts from the theoretical values. However these differences from the theoretical value should not be more than 5 mV for 10 in. × 10 in. recordings).

(4) A realistic value for n is 1, that is oxidation to the ferricinium ion; hence the wave is reversible and the formal electrode potential of the ferrocene/ferricinium couple in 0.1 M tetraethylammonium perchlorate in nitromethane is $+0.31$ V *versus* the aq. S.C.E.

It is sometimes asked what proportion of polarographic and voltammetric waves are reversible. Usually the questioner wishes to know whether reversible waves will result if polarography and voltammetry are applied to the types of compounds which interest him. The question is not easy to answer and perhaps about one third of all the polarographic and voltammetric waves which have been obtained are reversible or quasireversible. One investigator may find that every compound which he prepares produces a reversible oxidation or reduction wave, while another is constantly frustrated by obtaining a succession of very irreversible waves for his type of compounds.

There are two types of inorganic species which generally produce reversible waves. The first type consists of complexes where the starting material is oxidized or reduced to a structurally-similar ion or molecule; in such ions or molecules, the electron(s) is (are) added or removed from orbitals such that the bond lengths and angles scarcely change and such electron transfer processes are very fast (see Chapter 4). Compounds of this type are ferrocene, transition metal dipyridyl complexes and dithiolene complexes. This type of complex is frequently encountered by the preparative inorganic chemist, and the redox behaviour is soon elucidated using polarography and voltammetry.

The second type are the very labile complexes where the complexed metallic ions are reduced to the amalgams. Complexed or solvated ions of sodium, potassium, rubidium, caesium, barium, thallium (I), tin (II), lead (II), manganese (II), cadmium (II) and zinc (II) frequently fall into this class. These are the complexes whose stability constants, both in aqueous and nonaqueous solvents, are readily obtained polarographically.

Catalytic currents

Occasionally the inorganic chemist may come across a catalytic wave which results from the following mechanism:

$$O + ne \rightarrow R$$
$$R + C \rightarrow O$$

The oxidized form of the couple is regenerated by reaction of the reduced species with a species C, which is itself not reduced at the potentials being applied. The polarographic or voltammetric reduction of C is very irreversible and occurs to a negligible extent at the potentials being applied to the cathode, but chemical reaction between the reduced form of the couple and C takes place at a reasonable rate. In aqueous solution, catalytic waves are often obtained when R is a complex ion of a lower oxidation state of a transition element, such as molybdenum (III), and C is nitrate, perchlorate, hydrogen ion, hydrogen peroxide, and so on.

The possibility of a catalytic wave is always indicated by the chemistry of the system. The limiting current on a catalytic wave is made up of the normal diffusion current for the reducible species plus the catalytic current. As with kinetic waves, polarographic catalytic waves are such that the wave height is either independent of h_{corr}, or increases with h_{corr} but less rapidly than for an entirely diffusion-controlled wave. The formal electrode potential of the appropriate couple should not be equated with the half-wave potential of a catalytic wave.

Adsorption waves

Very occasionally the inorganic chemist may meet with an adsorption wave in addition to the normal diffusion wave, if the original material or its oxidation or reduction product is adsorbed on the electrode surface. Treatises on polarography should be consulted for further details, but if two waves are obtained, where the chemistry of the system is such that only one is to be expected, the second wave may be an adsorption wave.

Maxima

Although an S-shaped polarographic oxidation or reduction wave with a flat plateau is expected for an electroactive substance, quite frequently the current-voltage recording is not of this shape, but is distorted by the appearance of a maximum or even two maxima. Two types of maxima are known, referred to as those of the first or second kind. A maximum of the first kind is shown in Fig. 13; this type of maximum is superimposed on the rising part of a polarographic wave and makes the accurate determination of the half-wave potential of the wave impossible. At more negative potentials the current falls to the normal plateau current. Maxima of the second kind usually appear at fairly negative potentials and are frequently broader and shallower than the first kind. When maxima of the second kind appear, they are frequently on the plateaux of reduction waves.

Both kinds of maxima result from streaming of the solution past the surface of the drop. Although they have received extensive study, completely satisfactory explanations for their occurrence have not yet been found. Theories which are partly successful in explaining their presence have been published and details about these may be found in textbooks on polarography. Fortunately both kinds of maxima can usually be suppressed by adding minute quantities of surface-active agents to the solutions.

In aqueous solutions, Triton X–100, a non-ionic detergent, is the most widely used maximum suppressor. Meites [19] has suggested that the minimum concentration of Triton X–100 should be used to suppress a maximum, and that its concentration in the solution under examination should never exceed 0·004% (W/V); higher concentrations of Triton X–100 can have most undesirable effects on the wave itself. For non-aqueous solvents much less information is available on maximum suppressors, but methyl cellulose has been successfully employed to suppress some maxima in acetic acid–acetic anhydride mixtures.[30]

With irreversible waves, quasireversible waves, catalytic currents, adsorption waves and maxima, it might be thought that polarography and voltammetry are so fraught with inconveniences that the inorganic chemist is indeed

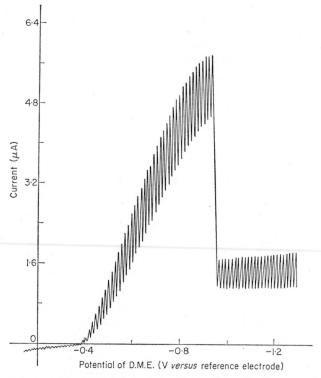

FIG. 13. A maximum of the first kind on the polarographic reduction wave for approximately 1.2×10^{-4} M bismuth (III) in acetic acid–acetic anhydride (20 : 1, V/V), which was 0.25 M in sodium acetate. The reference electrode was mercury-mercurous acetate–saturated lithium acetate in the organic solvent.

fortunate if he can quickly equate a half-wave potential with a formal electrode potential. However, this is not so for a great many systems and particularly for those complexes which are oxidized or reduced to structurally-similar ions or molecules. It is as well, however, to be aware of the difficulties which can arise in certain circumstances.

Determination of Stability Constants and Co-ordination Number for Complexes

If a metal ion and its complex reduce reversibly to the amalgam, then measurements of the shift in half-wave potential to more negative potentials with increase in ligand concentration can be applied to determine the overall stability constant and the number of ligands attached to the metal ion.

When a fairly stable complex exists in the solution, the complexing agent being in large excess:

$$(E_{\frac{1}{2}})_s - (E_{\frac{1}{2}})_c = \frac{0 \cdot 059}{n} \log \beta_p + \frac{0 \cdot 059}{n} p \log C \qquad \text{at } 25°C \qquad (3.6)$$

where $(E_{\frac{1}{2}})_s$ and $(E_{\frac{1}{2}})_c$ are the half-wave potentials for the simple and complexed ions respectively, p is the number of ligands coordinated to the metal ion, β_p is the overall stability constant and C is the ligand concentration.

A plot of $-(E_{\frac{1}{2}})_c$ against $\log C$ has a slope of $0 \cdot 059 \, p/n$ and since n is known, p can be found. Also at $C = 1$,

$$(E_{\frac{1}{2}})_s - (E_{\frac{1}{2}})_c = \frac{0 \cdot 059}{n} \log \beta_p. \qquad (3.7)$$

This method has been widely used to determine overall stability constants. It is necessary to maintain a constant ionic strength and, since the shifts in half-wave potential are only a few tenths of a volt, half-wave potentials must be measured to a tenth of a millivolt. Such accuracy is not possible on most recording polarographs and good manual polarographs are often employed. The above method should not be applied to quasireversible waves.

When the overall stability constant has a low value, then more than one complex will exist in solution, if $p \geqslant 2$. Under these conditions, a plot of $-(E_{\frac{1}{2}})_c$ against $\log C$ is a curve and the work necessary for extracting the stepwise stability constants is much more involved. The method is outlined by Crow and Westwood [46]; one system which has been examined by this latter treatment, for example, is the cadmium-thiocyanate system.[47] For solutions of ionic strength 2 M, $E_{\frac{1}{2}}$ changes from $-0 \cdot 5724$ V to $-0 \cdot 6646$ V over the thiocyanate concentration range of $0 \cdot 1$–$2 \cdot 0$ M. The complex species present in solution are $CdSCN^+$, $Cd(SCN)_2$, $Cd(SCN)_3^-$ and $Cd(SCN)_4^{2-}$ with stepwise stability constants of 11, 56, 6 and 60 respectively.

In certain cases, the stability constants and ligand numbers for complexes giving irreversible reduction waves can be obtained. Generally, methods involving competition for the ligand by the metal being studied and an indicator ion are used.[46]

CHAPTER 4

The Kinetics of Electron Transfer Reactions at an Electrode

WHEN a platinum electrode is immersed in a solution containing the oxidized and reduced forms of a couple, for example ferrocene and ferricinium ions, a dynamic system exists and the following electron transfer reactions are continuously taking place at the surface of the electrode:

$$[(\pi\text{-}C_5H_5)_2Fe]^+ + Pt \text{ with } e \rightarrow [(\pi\text{-}C_5H_5)_2Fe]^0 + Pt$$

$$[(\pi\text{-}C_5H_5)_2Fe]^0 + Pt \rightarrow [(\pi\text{-}C_5H_5)_2Fe]^+ + Pt \text{ with } e.$$

Under these conditions the potential of the electrode is the equilibrium electrode potential and at this potential there is no net flow of current in either direction across the electrode-solution boundary. There is, however, an appreciable current equal in both directions flowing across the boundary and this is the exchange current, i_{ex}.

The energy diagram for this and similar systems is shown in Fig. 14. It is assumed that the electron transfer reaction consists of only one rate-determining step which involves the same number of electrons as the overall electrode reaction.[48, 49] Consider the reduction of B to A with a rate of diffusion considerably faster than the rate of electron transfer at the equilibrium electrode potential. A and B are the potential energy curves for the reduced and oxidized forms, respectively; ΔG_{eq} is the free energy of activation at the equilibrium electrode potential. The cathodic current, which equals the exchange current, is proportional to $\exp(-\Delta G_{eq}^{\ddagger}/RT)$ where R is the gas constant and T the absolute temperature.

If the electrode potential is changed from E_{eq}, the equilibrium electrode potential, to a more negative potential by an amount ΔE so as to favour the cathodic process, that is the reaction B→A, the free energy of the system is changed by an amount $nF\Delta E$, where n is the number of electrons involved in the electron transfer process and F is the Faraday. This produces curve C. Now

$$\Delta G_c^{\ddagger} = \Delta G_{eq}^{\ddagger} - \alpha nF\Delta E \qquad (4.1)$$

where $\Delta G_c^{\ddagger}$ is the free energy of activation of the cathodic process and α is the transfer coefficient. The transfer coefficient is a measure of the symmetry of the energy barrier and is often about 0·5. For a consideration of the exact meaning and significance of the transfer coefficient, a recent article by Bauer [50] should be consulted. $\Delta G_c^{\ddagger}$ is now smaller than $\Delta G_{eq}^{\ddagger}$; hence the cathodic process occurs more readily and the cathodic current increases. Also

$$\Delta G_a^{\ddagger} = \Delta G_{eq}^{\ddagger} + (1-\alpha)nF\Delta E \qquad (4.2)$$

where ΔG_a is the free energy of activation of the anodic process. Clearly $\Delta G_a^{\ddagger}$ is now greater than $\Delta G_{eq}^{\ddagger}$. The anodic process occurs less readily and the anodic current decreases.

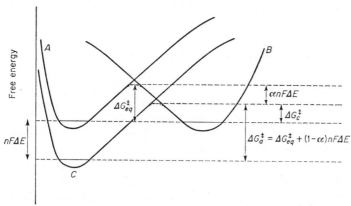

FIG. 14. Reaction coordinate diagram for an electron transfer reaction at an electrode. (Reprinted with permission from Kolthoff and Elving, 1963. "Treatise on Analytical Chemistry", Interscience, New York.)

The cathodic current at any potential E is actually given by

$$i_c = nFAk'C_{ox} \exp\left[-\alpha nF(E-E^{0'})/RT\right] \qquad (4.3)$$

where A is the area of the electrode, k' is the value of the heterogeneous rate constant at the formal electrode potential, $E^{0'}$, and C_{ox} is the concentration of the oxidized species at the electrode surface.

The anodic current is given by

$$i_a = -nFAk'C_{red} \exp\left[(1-\alpha)nF(E-E^{0'})/RT\right] \qquad (4.4)$$

where C_{red} is the concentration of the reduced species at the electrode surface. Therefore

$$i = i_c + i_a \qquad (4.5)$$

where i is the net current.

Now $\exp(-\alpha nF(E-E^{0'})/RT)$ increases as E becomes more negative; but $\exp[(1-\alpha)nF(E-E^{0'})/RT]$ decreases as E becomes more negative. Because of this, i_a can be neglected for a totally irreversible polarographic or voltammetric cathodic wave, that is a wave for which there is no detectable current at the formal electrode potential. Hence i equals i_c. At sufficiently negative potentials the cathodic current would become infinite according to eqn (4.3), but in deriving that equation the rate of diffusion was considered to be appreciably greater than the rate of electron transfer at the equilibrium electrode potential. The rate of diffusion is, of course, constant and finite, and does not allow the current to exceed a limiting value.

At the equilibrium electrode potential

$$i_{ex} = nFAk'C_{red}^{\alpha}C_{ox}^{1-\alpha}. \tag{4.6}$$

At the formal electrode potential

$$i_{ex} = nFAk'C \tag{4.7}$$

where C is the concentration of both the oxidized and reduced species at the electrode surface.

In Chapter 3 it was stated that the shape of a polarographic wave is readily derived by applying diffusion theory to the Nernst equation provided that the rate of the electron transfer reaction is considerably in excess of the rate of diffusion. Such a wave is reversible. In actual fact a polarographic wave is reversible when the value of the heterogeneous rate constant at the formal electrode potential exceeds 2×10^{-2} cm sec^{-1} at 25°C. If k' is less than this value, the wave is irreversible and becomes more irreversible as the value of k' decreases; if k' is less than 3×10^{-5} cm sec^{-1} the couple is totally irreversible.[51]

Values for heterogeneous rate constants at formal electrode potentials and for transfer coefficients are evaluated by electrical relaxation and steady-state methods.[49, 52] The relaxation methods involve measurements on chrono-amperograms obtained as a result of abruptly changing working electrode potentials from their equilibrium values, and on chronopotentiograms (see Chapter 5). They also include a.c. cell current and cell impedance measurements obtained after an a.c. voltage has been superimposed on the equilibrium potential of a working electrode. The steady-state methods are polarography and voltammetry, the kinetic parameters being obtained by "wave analysis".

Values of heterogeneous rate constants at the formal electrode potentials and transfer coefficients for a number of couples are shown in Table 2,[49] along with the effect of k' on the polarographic and voltammetric waves.

The Mechanism of Electron Transfer Reactions

As stated in Chapter 1, electrochemical techniques are attractive to inorganic chemists when, among other things, these techniques yield quick information on formal electrode potentials. When electron transfer reactions are fast and are free from accompanying slow chemical reactions, polarographic and voltammetric waves are reversible and formal electrode potentials are quickly obtained from half-wave potential data. Cyclic voltammograms and chronopotentiograms also yield quick information on formal electrode potentials when the electron transfer reactions are fast. Therefore, it is reasonable to enquire into what type of compounds will be oxidized and reduced rapidly at electrode surfaces, and if the rate constants for homogeneous electron transfer reactions involving a particular compound can help determine whether that compound will have a large heterogeneous rate constant at the electrode surface.

TABLE 2

Heterogenous Rate Constants, Transfer Coefficients and the Reversibility or Irreversibility of Polarographic and Voltammetric Waves

Reaction	Electrode	Base electrolyte	k' (cm sec^{-1})	α	Comments on polarographic and voltam-metric reduction waves
$Bi^{3+} \leftrightarrow Bi$	Bi(Hg)	1 M HClO$_4$	3×10^{-4}	—	Irreversible
$Bi^{3+} \leftrightarrow Bi$	Bi(Hg)	1 M HCl	>1	—	Reversible
$Cd^{2+} \leftrightarrow Cd$	Cd(Hg)	1 M KCl	2·9	0·78	Reversible
$Cr^{3+} \leftrightarrow Cr^{2+}$	Hg	1 M KCl	1×10^{-5}	—	Totally irreversible
$Cu^{2+} \leftrightarrow Cu$	Cu(Hg)	1 M KNO$_3$	$4·5 \times 10^{-2}$	—	Reversible
$Eu^{3+} \leftrightarrow Eu^{2+}$	Hg	1 M KCl	$2·1 \times 10^{-4}$	—	Irreversible
$Fe^{3+} \leftrightarrow Fe^{2+}$	Pt	1 M HClO$_4$	5×10^{-3}	—	Irreversible
$K^+ \leftrightarrow K$	K(Hg)	1 M Me$_4$N$^+$OH$^-$	0·1	—	Reversible
$Mn^{2+} \leftrightarrow Mn$	Hg	1 M KCl	3×10^{-3}	0·5	Irreversible
$Pb^{2+} \leftrightarrow Pb$	Pb(Hg)	1 M NaClO$_4$	3·3	0·6	Reversible
$Tl^+ \leftrightarrow Tl$	Tl(Hg)	1 M HClO$_4$	1·8	0·5	Reversible
$Zn^{2+} \leftrightarrow Zn$	Zn(Hg)	1 M KCl	4×10^{-3}	—	Irreversible

For many compounds, the answer to this question is in the affirmative and, since more information is available on homogeneous than on heterogeneous rate constants, a very brief account of the mechanism of homogeneous electron transfer reactions is appropriate. A vast amount of information, both theoretical and experimental, has been amassed in this field during the last twenty years and for detailed discussions on this topic other sources should be consulted.[53-57]

The homogeneous electron transfer reactions of principal interest for comparison with electron transfer reactions at electrodes are the homonuclear electron-transfer reactions such as

$$Fe(dipy)_3^{3+} + {}^*Fe(dipy)_3^{2+} \rightleftharpoons Fe(dipy)_3^{2+} + {}^*Fe(dipy)_3^{3+}.$$

These reactions are followed by isotopic labelling, nuclear magnetic resonance and electron spin resonance methods. They can proceed through an activated complex of either the outer-sphere type or the inner-sphere type. With the outer-sphere type, for example, the electron transfer reaction between cobaltocene and cobalticinium ions, the ligands bonded to the central metal atom remain attached to the atom during the process. The transfer of an electron takes place when these species come together to form the activated complex. A great many electron transfer reactions proceed by this mechanism and conclusive proof for this mechanism exists in many instances. This

TABLE 3

Rate Constants for the Removal of One Ligand from certain Transition Metal Complexes

Reactant	Rate constant (min^{-1})	Temp. $(^{\circ}C)$	References
$[\text{Fe(phen)}_3]^{3+}$	6×10^{-3}	25	58
$[\text{Fe(phen)}_3]^{2+}$	$4 \cdot 3 \times 10^{-3}$	25	59, 60
$[\text{Fe(dipy)}_3]^{2+}$	$7 \cdot 3 \times 10^{-3}$	25	61
$[\text{Fe(CN)}_6]^{4-}$	3×10^{-9}	20	62
$[\text{Ni(phen)}_3]^{2+}$	$4 \cdot 6 \times 10^{-4}$	25	63

TABLE 4

Rate Constants for some Homonuclear Electron Transfer Reactions with the Outer Sphere Mechanism

Reactants	Rate constant $(\text{l mole}^{-1} \text{ sec}^{-1})$	Temp. $(^{\circ}C)$	References
$^a[(\pi\text{-}C_5H_5)_2\text{Fe}]^{0,\,+}$	9×10^5	-75	64
$[\text{Fe(phen)}_3]^{2+,\,3+}$	10^5	25	65
$[\text{Os(dipy)}_3]^{2+,\,3+}$	5×10^4	25	65
$[\text{MnO}_4]^{2-,\,-}$	$3 \cdot 6 \times 10^3$	25	66
$[\text{IrCl}_6]^{3-,\,2-}$	10^3	25	67
$[\text{Fe(CN)}_6]^{4-,\,3-} (0\cdot01 \text{ M KOH})$	355	0	68

^{a} Methanol as solvent; elsewhere water.

mechanism must operate when the rate of dissociation of a ligand from the complex is much slower than the rate of electron transfer, and in Table 3 the rates of dissociation in water for the removal of one ligand are shown for a number of complex ions.

By comparison the rate constants of some homonuclear electron transfer reactions with the outer sphere mechanism are given in Table 4.

By selecting the *tris* (1,10-phenanthroline) iron (II) and iron (III) complexes from both tables, it is obvious that the electron transfer reaction is much faster than the ligand dissociation reactions, and there can be no doubt about the electrons transferring by an outer sphere mechanism.

The rate constants shown in Table 4 give the impression that electron transfer reactions *via* an outer sphere mechanism may always be fairly fast, but this is not so. For example, the rate constant for the electron exchange between *tris*-ethylenediaminecobalt (II) and *tris*-ethylenediaminecobalt (III) ions is low at 5×10^{-5} l mole^{-1} sec^{-1} at 25°C.[69]

When the ligands in complexes are labile conclusive proof of an outer sphere mechanism for homonuclear electron transfer reactions is more difficult to obtain, but it seems very likely that with these labile complexes electron transfers are also frequently achieved by an outer sphere mechanism.

For an electron transfer reaction by an outer sphere mechanism, the rate will be high if the activation energy associated with the formation of the activated complex is low; the activation energy will be low if the electron goes into or is taken from an orbital which does not have an appreciable influence on the bonds in the compound. Such facile electron transfers occur with compounds such as 2,2′-dipyridyl, 1,10-phenanthroline and di-thiolene complexes of transition metal ions, π-cyclopentadienyl and related metal complexes, substituted borazines, π-dicarbollyl transition metal complexes and certain cluster compounds.

The other mechanism for electron transfer involves an inner sphere type of activated complex. Here a ligand must dissociate from one of the complexes as the first step in the formation of the inner-sphere activated complex. This mechanism is also probably quite common, but conclusive proof has only been obtained for a few systems. The following electron transfer reaction illustrates this mechanism [70]:

$$[Co^{III}(NH_3)_5Cl]^{2+} + [Cr^{II}(H_2O)_6]^{2+} + 5H^+ \rightarrow [Cr^{III}(H_2O)_5Cl]^{2+} +$$
$$\underset{\text{inert}}{\qquad} \underset{\text{labile}}{\qquad} \underset{\text{inert}}{\qquad}$$
$$[Co(H_2O)_6]^{2+} + 5NH_4^+.$$
$$\underset{\text{labile}}{\qquad}$$

Note that the chloride ion has been transferred from the cobalt atom to the chromium atom and this can only occur if the following bridged activated complex is formed:

$$[Cr(H_2O)_6]^{2+} \rightarrow [Cr(H_2O)_5]^{2+} + H_2O$$
$$[Co(NH_3)_5Cl]^{2+} + [Cr(H_2O)_5]^{2+} \rightarrow [(NH_3)_5Co\ Cl\ Cr(H_2O)_5]^{4+}.$$

Like electron transfers by the outer sphere mechanism, the rate constants for electron transfers by the inner sphere mechanism are also very variable. Some data to illustrate this point are shown in Table 5[71] for reactions similar to that above.

This inner sphere mechanism is also probably operative in some homo-nuclear electron transfer reactions.

TABLE 5

Kinetic Data for Cr^{2+}—$Cr(NH_3)_5X^{2+}$ Reactions at 25°C

Reactants	Rate constant ($l\,mole^{-1}\,sec^{-1}$)
$Cr^{2+}+[Cr(NH_3)_5F]^{2+}$	$2\cdot7\times10^{-4}$
$Cr^{2+}+[Cr(NH_3)_5Cl]^{2+}$	$5\cdot1\times10^{-2}$
$Cr^{2+}+[Cr(NH_3)_5Br]^{2+}$	$0\cdot32$
$Cr^{2+}+[Cr(NH_3)_5I]^{2+}$	$5\cdot5$

Homonuclear and Electrochemical Rate Constants

For one-electron transfers occurring by an outer sphere mechanism, Marcus [72, 73] has predicted that the following relationship will often apply.

$$(k_{ex}/Z_{soln})^{\frac{1}{2}} \simeq k_{el}/Z_{el} \tag{4.8}$$

where k_{ex} and k_{el} are the homonuclear and heterogeneous rate constants respectively, and Z_{soln} and Z_{el} are collision frequencies, namely about 10^{11} cc $mole^{-1}\,sec^{-1}$ and 10^4 cm sec^{-1}, respectively.

Comparisons made between values calculated from measured rate constants are shown in Table 6 and on the whole the agreement is quite good.[27] k_{el} is the heterogeneous rate constant at the formal electrode potential, referred to earlier as k′.

TABLE 6

A Comparison of Kinetic Data obtained from Homonuclear and Electrochemical Rate Constants

System	Medium	$(k_{ex}/10^{11})^{\frac{1}{2}}$	$k_{el}/10^4$	Electrode
$[Fe(CN)_6]^{4-,\,3-}$	1 M K^+	1×10^{-3}	1×10^{-5}	Pt
$[MnO_4]^{2-,\,-}$	0·9 M Na^+	2×10^{-4}	$\sim2\times10^{-5}$	Pt
$Fe^{2+,\,3+}$	1M $HClO_4$	9×10^{-6}	7×10^{-7}	Pt
$V^{2+,\,3+}$	1M $HClO_4$	4×10^{-7}	4×10^{-7}	Hg
$Eu^{2+,\,3+}$	1 M Cl^-	6×10^{-8}	3×10^{-8}	Hg
$[Co(NH_3)_6]^{2+,\,3+}$	0·14 M H^+	$<5\times10^{-11}$	$\sim5\times10^{-12}$	Hg

Since a polarographic wave is reversible when the value of the heterogeneous rate constant exceeds 2×10^{-2} cm sec^{-1} at 25°C, it can therefore be predicted that, for outer sphere mechanisms, reversible polarographic waves are likely if the rate constant for the homonuclear electron-transfer reaction in solution exceeds 4×10^{-1} l $mole^{-1}\,sec^{-1}$ at 25°C. For a great many couples this value is exceeded.

In the case of the polarographic reduction of solvated or other complex ions to the amalgam, the Marcus theory cannot apply. In these cases, electrons from the mercury could be transferred before the ligands dissociate or after one or more ligands have dissociated. For the electrochemical system to be completely reversible, these accompanying chemical reactions must, like the electron transfer reactions, be fast compared with the rate of diffusion of the complex ions to the mercury surface.[12]

CHAPTER 5

Cyclic Voltammetry, Chronopotentiometry and other Techniques for Verifying Reversibility

Cyclic Voltammetry

As HAS been stressed in Chapter 3, the formal electrode potential of a reversible couple is to a very good approximation equal to the identical half-wave potentials of the reduction wave for the oxidized form and the oxidation wave of the reduced form. However, both forms of the couple are seldom available and, if there is doubt about the reversibility of a couple after a polarographic or voltammetric investigation because, for example, a relatively slow chemical reaction following a fast electron transfer reaction is not ruled out by the chemistry of the system, then a quick test of reversibility is desirable. Cyclic voltammetry makes this possible.

In this technique, the most convenient working electrode if the polarographic wave is found at a potential more negative than the potential of the oxidation of mercury is a hanging mercury drop, or, if this is not so, it is the cylindrical micro-platinum electrode used in voltammetry as a rotating platinum electrode. The cells are identical to those used in polarography and voltammetry and are maintained at a constant temperature, usually 25°C. The information given in Chapter 3 about solvents, base electrolytes, concentrations of electroactive species, working ranges, the removal of oxygen and correction for potential drop across the cell, if a controlled potential polarograph is not used, is equally relevant to cyclic voltammetry.

The construction of a cylindrical micro-platinum electrode was discussed in Chapter 3; a hanging mercury drop electrode can be made in the following way.[74] A short piece of platinum wire is sealed in the end of a length of glass tubing, the protruding end of wire is cut off close to the glass and the wire is filed flush with the glass surface. Any oxide film on the platinum surface is removed with acidified ferrous sulphate solution and the platinum plated with mercury, by making it the cathode in a cell with a mercury pool anode and an electrolyte of concentrated mercuric nitrate in 1 M nitric acid. A potential of 1·5 V across the cell with a plating time of 5 min is used. Mercury drops will adhere satisfactorily to the plated surface; one or two drops of mercury are collected on a Teflon scoop from a conventional dropping mercury electrode immersed in the solution, and these are then transferred in the solution to the plated surface of the new electrode, which itself is immersed in the solution. The mercury in the scoop adheres to the mercury surface to produce a stationary mercury drop.

With cyclic voltammetry the working electrode is suspended in a still solution and the potential of the working electrode is changed at a constant rate backwards and forwards between two limits as shown in Fig. 15. These limits should lie within the polarographic or voltammetric range for the solution under study. If the system is such that only one d.c. polarographic or voltammetric wave is to be expected, then the range over which the potential is scanned can be as small as 0·5 V and should be approximately centred on the d.c. polarographic or voltammetric wave. The current, which flows through the cell, is recorded as a function of the potential of the working electrode, as in d.c. polarography, but here the working electrode is stationary.

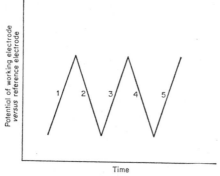

Fig. 15. The potential of the working electrode as a function of time in cyclic voltammetry.

When only one potential sweep is made (1 in Fig. 15), the recording of current against applied potential is called a single-sweep voltammogram. A typical single-sweep voltammogram is shown in Fig. 16.[75] Note that the current–potential recording is peaked. In voltammetry with a rotating platinum electrode no peak is obtained because the solution is stirred and the diffusion layer is of constant thickness for potentials on the plateau of the voltammogram. However with a stationary electrode the solution is quiescent and the diffusion layer continues to increase in thickness as time passes at potentials where the electroactive substance is oxidized or reduced. In Fig. 16, all the thallous ions reaching the electrode surface after the peak potential are immediately reduced, but because the diffusion layer is thickening, the supply of thallous ions to the electrode decreases and the current falls.

It is not difficult to derive the shape of the current–potential curve for a reversible oxidation or reduction wave in polarography, but to obtain theoretically the shape of a single-sweep voltammogram demands more complex mathematics. However, for reversible systems such theoretical current–potential curves have been derived for linear diffusion to a shielded plane electrode,[76, 77] for cylindrical diffusion to a wire electrode,[78] and for spherical diffusion to a hanging mercury drop.[75]

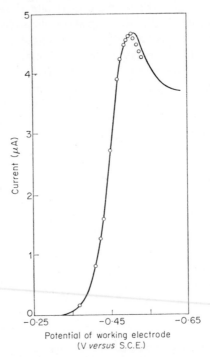

FIG. 16. A single sweep voltammogram for the reduction of 1·00 mM thallium (I) in 0·1 M aqueous potassium chloride. The working electrode was a hanging mercury drop and the rate of voltage change was 0·01389 V/sec^{-1}. The line is experimental; the points are theoretical. (Reprinted with permission from *J. Am. chem. Soc.*, **78**, 2972, 1956.)

For linear diffusion where a species in solution is reduced or oxidized to a species in solution [15]

$$i_p = 2{\cdot}687 \times 10^5 \, . \, n^{\frac{3}{2}} A D^{\frac{1}{2}} C v^{\frac{1}{2}} \tag{5.1}$$

where i_p = peak current (μA); n = number of electrons involved in the oxidation or reduction; A = area of the electrode (cm^2); D = diffusion coefficient of the electroactive species (cm^2 sec^{-1}); C = concentration of the electroactive species in the solution (mmoles l^{-1}); and v = sweep rate (V sec^{-1}).

For cylindrical and spherical diffusion, similar equations have been derived but the peak currents are slightly larger in these cases for constant n, A, D, C and v because more electro-active species can approach a definite area of the electrode in a given period of time. For the conditions usually employed in practical work with the hanging mercury drop and the micro-platinum wire electrode,

$$E_p = E_{\frac{1}{2}} - \frac{0 \cdot 0285}{n} \text{ (volts)} \tag{5.2}$$

for a reversible reduction. E_p is the potential at which the peak occurs and $E_{\frac{1}{2}}$ is the polarographic half-wave potential. For an oxidation wave

$$E_p = E_{\frac{1}{2}} + \frac{0 \cdot 0285}{n} \text{ (volts)} \tag{5.3}$$

For an irreversible reduction, the rise in current to the peak is less steep and the peak potential is more negative than that to be expected for a reversible reduction.

It is, however, the combination of a forward single-sweep with a backward sweep which is of primary interest to the inorganic chemist, because the species reduced on the forward sweep is usually oxidized on the backward sweep and another peaked voltammogram is produced. A typical cyclic voltammogram showing one forward sweep and one reverse sweep is shown in Fig. 17 for a reversible system. For such systems, where the fast electron transfer reaction is unaccompanied by slow chemical reactions, the heights of the cathodic and anodic peaks are approximately equal.

For a reversible system, the peak potential of a reduction wave is $28 \cdot 5/n$ mV more negative than the corresponding half-wave potential, and the peak potential of an oxidation wave is $28 \cdot 5/n$ mV more positive than the corresponding half-wave potential. For such a system, the cyclic voltammogram has the same general shape irrespective of which form of the couple is in the solution at the start of the first potential sweep. The separation of the peak potentials for a reversible couple is, therefore, $57/n$ mV and the formal electrode potential of the couple is the average of the two peak potentials to a good approximation. If the electron transfer reaction is irreversible, then the peak-to-peak distance is increased.

Of course, one can continue to cycle between the selected limits of potential of the working electrode and first, second, third, fourth, and so on, peaks are obtained. For a reversible system the peak potentials of the first, third, fifth, and so on, peaks (forward sweeps) are identical; so are the peak potentials for the second, fourth, sixth, and so on, peaks (backward sweeps). Only two sweeps are required to test the reversibility of a system.

The sweep rate can vary from, say, 1 V per 0·01 sec to 1 V per 100 sec but slower sweep rates than this latter value should not be used, for at such sweep rates it is difficult to prevent some convective mixing in the diffusion layer. For sweep rates faster than 1 V per 5 sec an oscilloscope must be used to record the current–potential curves, but a potentiometric recorder with a fast response to changes in potential can be used at slower sweep rates. The author uses a Beckman Electroscan 30 in his work, where the direction of rotation of the motor driving the applied potential drum and the chart paper is reversed manually at the selected potential limits. With this instrument, sweep rates of 1 V per 5–100 sec are readily obtained, and d.c. polarograms and chronopotentiograms are obtained with the same instrument. However, it is advisable to check the actual peak-to-peak separation on a system

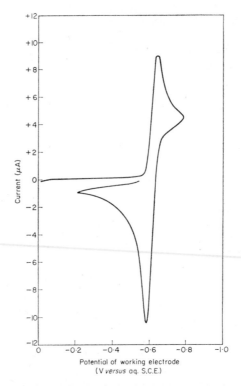

FIG. 17. A cyclic voltammogram for 10^{-3} M cadmium perchlorate in dimethylformamide, which was 0·1 M in tetraethylammonium perchlorate. The working electrode was a hanging mercury drop and the rate of voltage change was 0·04 V/sec^{-1}.

known to be reversible with the equipment selected for cyclic voltammetry. If the sweep rate is too fast with a potentiometric recorder, there is a recorder lag and the peak separation for a reversible system is greater than $57/n$ mV. With slower sweep rates, there may be slight convective mixing in the diffusion layer and the peak potential on a reduction sweep will be slightly more negative than the theoretical peak potential, and *vice versa* for an oxidation sweep. This again leads to a peak separation in excess of $57/n$ mV. In Fig. 16 it will be noticed that the experimental peak potential is about 15 mV later than the theoretical peak potential, and Frankenthal and Shain [75] state that convective mixing is more likely with spherical diffusion (hanging mercury drop) than with cylindrical diffusion (micro-platinum wire). If, in a cyclic voltammogram, the peak potential for the oxidation sweep is also 15 mV late, then the peak separation for this reversible system is about 90 mV rather than 57 mV.

It follows that it must be ascertained what the experimental peak separations will be for a particular assembly and reversible systems of different n values. If the peak separation with another system exceeds the experimental value obtained for a reversible couple with the same n value, then the system is irreversible. The average of the peak potentials for the reduction and oxidation waves of a system must not be equated with the formal electrode potential of the couple under study, if the separation in peak potentials is appreciably greater than that obtained for a reversible system with the same n value. Of course, every effort must be made to keep the electrode stationary and the solution still by eliminating any vibrations in the equipment.

The reversing potential sweeps can also be applied with a triangular wave generator such as the Hewlett-Packard Function Generator Model 202A, and this allows rapid sweep rates to be applied to the electrodes in conjunction with an oscilloscope. However, if the sweep rates are very rapid, a couple which is reversible at slow sweep rates (say, 1 V per 20 sec) can become irreversible,[79] and this is a disadvantage for the inorganic chemist who wants quick information on the formal electrode potential of the couple.

Cyclic voltammetry is particularly useful for establishing that a polarographic or voltammetric wave is reversible and not quasireversible, and that the value of the formal electrode potential obtained from the wave cannot be questioned. If the chemistry of the system is such that a relatively slow chemical reaction could possibly precede the fast electron transfer, a simple test to settle this point has already been described on page 23. However, it is not possible from conventional polarographic or voltammetric data to recognize when a fast electron transfer is followed by a relatively slow reversible chemical reaction or an irreversible chemical reaction. A cyclic voltammogram will show at once if this type of coupled chemical reaction is present, because, if it is present, the height of the anodic peak is either appreciably less than the height of the cathodic peak or is completely absent, when the starting material is the oxidized form of the couple; and *vice versa* for the reduced form.[15] This latter type of quasireversible polarographic or voltammetric wave is quite common for the reduction of certain types of organometallic compounds, where the reduced species decomposes rapidly in solution. A relatively slow chemical reaction which precedes a rapid electron transfer reaction can also be recognized from a cyclic voltammogram, but the original paper should be consulted for details.[15]

Cyclic voltammetry has not been widely used by inorganic chemists, but the technique has been employed to investigate the reversibility of electron transfer reactions of transition metals dithiolene [3] and π-dicarbollyl [6] complexes, organometallic compounds,[80, 81] substituted borazines [1] and iron cyanide—BX_3 adducts.[2] Without doubt, cyclic voltammetry should be used more by inorganic chemists as the necessary equipment is not expensive. In fact for many reversible systems where n is known there is no need to conduct d.c. polarography or voltammetry as well as cyclic voltammetry; a cyclic voltammogram will give all the information that is required.

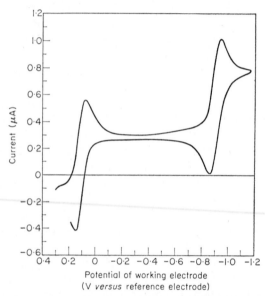

FIG. 18. Cyclic voltammogram of a 1 mM solution of $Ni[S_2C_2(C_6H_5)_2]_2$ in dimethylformamide, which was 0·1 M in lithium perchlorate. The reference electrode was silver–silver chloride–0·1 M aqueous lithium chloride. A platinum working electrode was used. (Reprinted with permission from *J. Am. chem. Soc.*, **88**, 4877, 1966.)

When two electroactive substances are in solution or the electroactive substance is reduced or oxidized in two steps, two peaks are obtained on the reduction and oxidation half-cycles. The inorganic chemist is more likely to meet the case of a substance being oxidized or reduced in two or more steps and, for a reversible system of this type, the peak currents on any sweep suitably corrected for background current will be of equal height if the n values are the same for each electron transfer step. This is illustrated for the reduction of the nickel dithiolene complex, $Ni[S_2C_2(C_6H_5)_2]_2$, in Fig. 18; here n is one for each electron transfer step.[3]

Most inorganic chemists will wish to use cyclic voltammetry as a rapid instrumental technique to obtain information on formal electrode potentials in the same way as they use infrared and n.m.r. spectroscopy for quick structural information, but it should be recognized that, like polarography and chronopotentiometry (see the next section), it is now a powerful technique for studying irreversible electron transfer processes, electron transfer reactions preceded or followed by relatively slow chemical reactions, and adsorption effects at electrodes.[15, 79] This use of cyclic voltammetry together with polarography, controlled potential coulometry, ultraviolet, infrared and e.s.r. spectrometry has been demonstrated by Rupp and co-workers [82, 83] in establishing the mechanism of the electrolytic reduction of decaborane (14) in 1,2-dimethoxyethane. The author refers inorganic chemists interested in using cyclic voltammetry to gain insight into complex redox reaction mechanisms to these papers.

Cyclic voltammetry in conjunction with voltammetry, controlled potential electrolysis and coulometry, and chronopotentiometry has also been used to study the kinetics of the electrochemical oxidative coupling of decahydro-clavodecaborate $(2-)$, $B_{10}H_{10}^{2-}$, in acetonitrile.[84]

Chronopotentiometry

In this technique a constant current is passed between the working and auxiliary electrodes in a suitable cell containing a still solution of a base electrolyte and the electroactive species; the potential of the working electrode *versus* a reference electrode is recorded as a function of time. From the chrono-potentiogram thus obtained, the formal electrode potential for the appropriate couple, of which one form is the electroactive substance, is readily obtained provided that the oxidation or reduction of the electroactive species is reversible. A diagram of the apparatus is shown in Fig. 19.

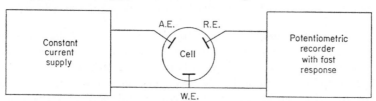

FIG. 19. Apparatus for chronopotentiometry. A.E., R.E. and W.E. are the auxiliary, reference and working electrodes respectively.

A continuously variable constant current supply should be used. The author employs a Beckman Electroscan 30, with which constant currents between 0 and 100 mA are available. A suitable cell for chronopotentiometry with a mercury working electrode and non-aqueous solvents is shown in Fig. 20.

Compartments B and C contain the base electrolyte. Compartment A contains the base electrolyte and electroactive substance. With aqueous solutions, compartment B is not required and the sinter at the end of the aqueous bridge is positioned just above the mercury surface. Similar cells can be designed for platinum electrodes. The areas of working electrodes are usually 1–5 cm^2.

If more than one chronopotentiogram is to be recorded, the solution in compartment A should be stirred after each recording and allowed to become quiescent before the next recording is made. As in polarography and voltammetry, the temperature of the solution should be kept constant, usually at

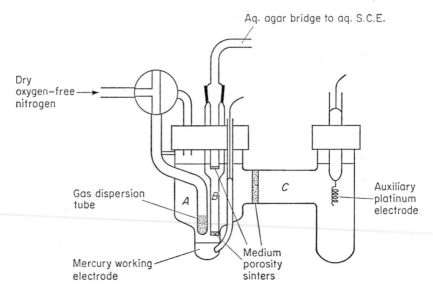

FIG. 20. A suitable cell for chronopotentiometry with non-aqueous solutions.

25°C, and oxygen should be removed from the solution by bubbling oxygen-free nitrogen before recording a chronopotentiogram, if the potential range is such that interference from oxygen is to be expected. Also, as in polarography, the concentration of the base electrolyte should be at least fifty times that of the electroactive species. A typical chronopotentiogram for a reversible process is shown in Fig. 21.[85]

Before a steady current is passed, the platinum working electrode for the system referred to in Fig. 21 is in contact with the reduced form of the couple and, say, a minute trace of the oxidized form. As a steady current is passed between the working and auxiliary electrodes, molecules of the reduced form of the couple at the surface of the working electrode are oxidized and the ratio of the oxidized to reduced species at the electrode surface changes quite rapidly. Hence the initial move to more positive potentials in Fig. 21 as dictated by the Nernst equation. However, when the concentrations of oxidized and reduced species are of the same order, the potential remains almost constant, only becoming more positive very slowly, which is to be expected from the Nernst equation. After a short time, however, practically all molecules of the reduced species reaching the electrode surface by diffusion are oxidized and the ratio of oxidized to reduced species changes rapidly, with an accompanying rapid change in the working electrode potential in a positive direction. The time from switching on the constant current until this rapid change in potential is achieved, is called the transition time. The potential of the working electrode does not, of course, keep moving rapidly in a positive direction, because eventually the potential is such that an ion of the base electrolyte or the solvent itself is oxidized. This causes the working electrode to assume, once more, a steady potential.

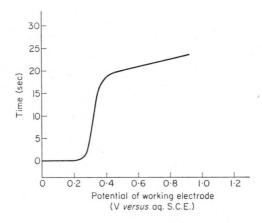

FIG. 21. Chronopotentiogram for 8.09×10^{-4} M methylferrocenylcarbinol in acetonitrile, which was 0·2 M in lithium perchlorate. The current density was 63·5 μA cm^{-2}. (Reprinted in part with permission from *J. Am. chem. Soc.*, **82**, 5812, 1960.)[85]

Using diffusion theory, it can be readily shown for linear diffusion that

$$\tau^{\frac{1}{2}} = \frac{\pi^{\frac{1}{2}} n F A D^{\frac{1}{2}} C}{2i} \tag{5.4}$$

where τ = transition time (sec); C = concentration of electroactive material (mmoles l^{-1}); i = constant electrolysis current (μA); D = diffusion coefficient (cm^2 sec^{-1}); F = the Faraday; A = electrode area (cm^2); and n = the number of electrons involved in the oxidation or reduction.

For a reversible reduction, where the reduced species is in solution,

$$E_{\text{w.e.}} = E^{0\prime} - \frac{0.059}{n} \log \frac{D_O^{\frac{1}{2}}}{D_R^{\frac{1}{2}}} - \frac{0.059}{n} \log \frac{t^{\frac{1}{2}}}{\tau^{\frac{1}{2}} - t^{\frac{1}{2}}} \qquad \text{at 25°C} \tag{5.5}$$

where t is the time of passage of the constant current and $E_{w.e.}$ is the potential of the working electrode after time t. When $t = \tau/4$, then $t^{\frac{1}{2}}/(\tau^{\frac{1}{2}} - t^{\frac{1}{2}}) = 1$ and

$$E_{\tau/4} = E^{0\prime} - \frac{0 \cdot 059}{n} \log \frac{D_0^{\frac{1}{2}}}{D_R^{\frac{1}{2}}} = E_{\frac{1}{2}}$$

for the corresponding reversible polarographic reduction wave. $E_{\tau/4}$ is called the quarter transition time potential. The term $0 \cdot 059/n \log D_0^{\frac{1}{2}}/D_R^{\frac{1}{2}}$ is approximately zero; hence

$$E_{\tau/4} = E_{\frac{1}{2}} \simeq E^{0\prime}. \tag{5.6}$$

Chronopotentiometry is, therefore, another means of determining the formal electrode potential for a reversible couple.

For a reversible reduction

$$E_{w.e.} = E_{\tau/4} - \frac{0 \cdot 059}{n} \log \frac{t^{\frac{1}{2}}}{\tau^{\frac{1}{2}} - t^{\frac{1}{2}}} \tag{5.7}$$

and a plot of $E_{w.e.}$ against $\log [t^{\frac{1}{2}}/(\tau^{\frac{1}{2}} - t^{\frac{1}{2}})]$ is a straight line of slope $-59/n$ mV. It can also be readily shown that $E_{3\tau/4} - E_{\tau/4}$ is $-48/n$ mV for a reversible reduction. The same equations hold for oxidations except that the sign before the term $0 \cdot 059/n \log t^{\frac{1}{2}}/(\tau^{\frac{1}{2}} - t^{\frac{1}{2}})$ is now positive.

In polarography and voltammetry, the half-wave potential can only be equated with the formal electrode potential when the reduction or oxidation of the electroactive species is reversible, with no relatively slow chemical reactions before or after the electron transfer step. The same applies in chronopotentiometry when equating a quarter transition time potential with a formal electrode potential. With an irreversible chronopotentiogram the change in potential in the vicinity of the transition time is less abrupt than for a reversible chronopotentiometric wave and, for a reduction, the quarter transition time potential is more negative than the formal electrode potential. Like irreversible polarograms and voltammograms, irreversible chronopotentiograms can be analysed to give data on transfer coefficients and heterogeneous rate constants. Also chronopotentiometry can be used to study electron transfer reactions preceded or followed by chemical reactions.[14]

For a reversible reduction to an insoluble product of unit activity, (5.5) does not apply but is replaced by the following equation.

$$E_{w.e.} = E^{0\prime} + \frac{0 \cdot 059}{n} \log \frac{C}{\tau^{\frac{1}{2}}} + \frac{0 \cdot 059}{n} \log (\tau^{\frac{1}{2}} - t^{\frac{1}{2}}) \qquad \text{at } 25\,^{\circ}\text{C} \tag{5.8}$$

where C is the bulk concentration of the reducible ion in the solution.

Although chronopotentiometry is very useful for investigating the kinetics of electron transfer reactions, it has been used much less by inorganic chemists than polarography and voltammetry for determining formal electrode potentials. This is probably because polarographs are available in most chemical laboratories, since polarography is a popular analytical technique.

Chronopotentiometry can certainly be used for quantitative analysis, since the transition time is proportional to the square of the concentration of a particular electroactive species, if the same electrode and same constant current are employed (see Eqn 5.4). However, as an analytical technique, it suffers from the disadvantage that the transition time for the second reduction in a system, where two species reduce at different potentials or the same species reduces in steps at different potentials, is dependent on the first transition time.

For mixtures of two substances, the first transition time is given by

$$\tau_1^{\frac{1}{2}} = \frac{\pi^{\frac{1}{2}} n_1 F A D_1^{\frac{1}{2}} C_1}{2i} \tag{5.9}$$

but the second transition time obeys the equation

$$(\tau_1 + \tau_2)^{\frac{1}{2}} - \tau_1^{\frac{1}{2}} = \frac{\pi^{\frac{1}{2}} n_2 F A D_2^{\frac{1}{2}} C_2}{2i}. \tag{5.10}$$

For the special case where a species is reduced in two steps and the number of electrons involved are n_1 and n_2,

$$(\tau_1 + \tau_2)/\tau_1 = (n_1 + n_2)^2 / n_1^2. \tag{5.11}$$

In polarography and voltammetry with a mixture of two electroactive materials, however, the height of the second wave, which is proportional to the concentration of the second electroactive substance, is independent of the height of the first wave. Hence the preference for polarography as an analytical technique. The phenomenon of chronopotentiometry expressed by Eqn (5.10) should not appreciably effect the usefulness of the technique for the inorganic chemist, but as a method for determining formal electrode potentials, chronopotentiometry is less satisfactory than polarography or voltammetry if the concentration of the electroactive substance is very low. Satisfactory polarographic or voltammetric waves can be obtained at concentrations of the electroactive substance equal to or in excess of 10^{-5} M. With chronopotentiometry the concentration of this substance should generally be in excess of 5×10^{-4} M.

In contrast to this disadvantage, however, chronopotentiometry has the distinct advantage over normal d.c. polarography and voltammetry in that the complete reversibility of a system can be readily checked by chronopotentiometry followed by current-reversal chronopotentiometry. If the direction of the constant current is reversed at the transition time, τ_f, then a second transition time, τ_r, results from the reverse reaction and $\tau_r = \tau_f/3$ provided that both the oxidized and reduced species are in solution and that there are no relatively slow or irreversible chemical reactions coupled with the electron transfer process. For the current-reversal chronopotentiogram, $E_{0.22\tau(r)}$ equals $E_{0.25\tau(f)}$ and both of these are, to a very good approximation, equal to the formal electrode potential of the appropriate couple, if it is reversible. Note that $E_{0.22\tau(r)}$ not $E_{0.25\tau(r)}$, is equal to the quarter transition time potential for the forward reaction, although the difference between $E_{0.22\tau(r)}$, and $E_{0.25\tau(r)}$, is actually almost negligible; it is $3.8/n$ mV.

It is asked whether the inorganic chemist should use polarography (or voltammetry), or cyclic voltammetry, or chronopotentiometry to determine formal electrode potentials. If the complete reversibility of the system under investigation is not in doubt, but the substance to be used is in short supply and very little can be spared for an electrochemical investigation, then polarography (or voltammetry) or cyclic voltammetry should be employed with, say, a 5×10^{-5} M solution of the electroactive material. If the complete reversibility of the system is in doubt, use cyclic voltammetry. If the substance under investigation is not in short supply, chronopotentiometry may be used as an alternative technique to determine the formal electrode potential, and the complete reversibility of the system may be checked using reversed current chronopotentiometry, if this is felt to be necessary. At the present time, normal d.c. polarography is by far the most popular technique for determining formal electrode potentials if only one form of the couple is available, or for obtaining some information about n values. However it seems likely that cyclic voltammetry and, perhaps, chronopotentiometry, will be used much more in the future by inorganic chemists, for it is certainly desirable to know beyond all doubt when one is dealing with a completely reversible rather than a quasireversible system.

Finally a few further practical details about chronopotentiometry are presented. Since the basic equations of polarography, voltammetry and chronopotentiometry are derived from diffusion theory and the Nernst equation, the information about useful working ranges, reference electrodes, base electrolytes, and so on, given in Chapter 3 is equally relevant in this chapter. Also, in polarography and voltammetry, wave heights have to be corrected for residual current, usually consisting mainly of the condenser current. For the most precise results in chronopotentiometry, transition times should be corrected for the residual transition time, that is the transition time found for the base electrolyte alone. With potentiometric recorders the transition time should exceed 5 sec but the value of the constant current should be such that the transition time does not exceed 100 sec because with the larger transition times, convective mixing may occur in the diffusion layer.

One interesting and useful application of chronopotentiometry is the determination of n values using very thin layers of electrolyte ($< 50 \mu$) in contact with the working electrode. Under these special conditions, a knowledge of the diffusion coefficient is not required, the relevant equation being

$$i\tau = nFAlC \tag{5.12}$$

where i = constant electrolysis current (A); τ = transition time (sec); n = number of electrons involved in the oxidation or reduction; F = the Faraday; A = electrode area (cm^2); l = layer thickness (cm); and C = initial concentration (moles cm^{-3}).

A convenient dip-type thin-layer electrolysis cell for the determination of n values in non-aqueous solvents is described by McClure and Maricle.[86] N values were determined for electroactive substances in acetonitrile, dimethylformamide and propylene carbonate to $\pm 4\%$.

For further details on chronopotentiometry, the reader is referred to articles by Delahay,[87] Lingane,[88] Meites[89] and Paunovic.[90]

Other Techniques

The reversibility of a couple, when only one form of the couple is available, can also be verified by large amplitude a.c. oscillographic polarography and voltammetry, in which an alternating current of frequency about 50 cycles

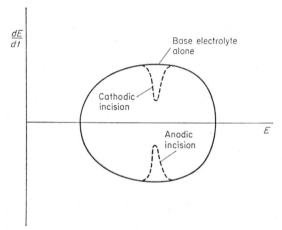

FIG. 22. An oscillopolarogram for the base electrolyte alone and with one of the species, which form of reversible couple. (Reprinted with permission from Meites.[19])

sec^{-1} and amplitude of 100–2000 μA is applied to a polarographic cell with the usual type of reference electrode and a dropping mercury electrode, hanging mercury drop or microplatinum wire as working electrode. A record of dE/dt against E, where E is the potential of the working electrode *versus* the reference electrode, is obtained on an oscilloscope. For the base electrolyte alone, an elliptical plot is obtained as shown in Fig. 22.[91]

When an electroactive species is present in the solution, incisions are produced on this ellipse, one corresponding to the reduction step and the other to the oxidation step (see Fig. 22). If the couple is reversible, the peak potentials of the incisions are practically identical and nearly equal to the polarographic halfwave potentials. For an irreversible couple the peak potentials are further apart. If equipment for oscillographic polarography is available, it can be used to verify the reversibility of a couple, but the method has no advantages over cyclic voltammetry for this purpose. For further details about this technique, articles by Meites [91] and Kaldova [92] should be consulted. Like cyclic voltammetry and chronopotentiometry, oscillographic polarography is a useful technique for studying the detailed mechanism of electrode reactions.

Alternating current polarography [93, 94] and the Kalousek commutator [95] can also be used for obtaining information about the reversibility of couples, but equipment for these techniques is less readily available than for cyclic voltammetry, chronopotentiometry and oscillographic polarography.

CHAPTER 6

Controlled Potential Electrolysis and Coulometry

IN THE previous̄chapters, electrochemical techniques have been described for obtaining information quickly about the formal electrode potentials of couples. Using polarography or voltammetry, it has also been shown how some information on n values can be obtained; one of the best techniques for determining n values, however, is controlled potential coulometry and this is described in this chapter. Once again emphasis will be given only to electron transfer processes which are uncomplicated by accompanying slow chemical reactions. As has been mentioned already, meaningful information about formal electrode potentials and n values can be obtained quickly if the electron transfer processes are fast and uncomplicated by slow chemical reactions. Many such systems have been investigated in recent years and, if the inorganic chemist thinks that his systems are in this class, then electrochemical techniques should be used to obtain formal electrode potentials and n values, just as infrared, ultraviolet, visible, n.m.r. and e.s.r. spectroscopy, and mass spectrometry are employed to obtain information on structure.

The author feels that most inorganic chemists, whose main interests are synthesis, structure and bonding, are unlikely to become engrossed in complicated electron transfer reactions whose complete understanding may take weeks of painstaking effort. The elucidation of the more complex electron transfer reactions is a fascinating and worthwhile field of study for the specialist and controlled potential coulometry is another most useful technique for this work [16, 22, 96] but only the simplest processes will be considered in this chapter.

In controlled potential electrolysis the potential of the working electrode is held at a fixed potential on the plateau of a polarographic or voltammetric wave, and the electroactive species is oxidized or reduced completely. Obviously the electrode has to be of a large surface area and the solution must be well stirred if the reaction is to be practically completed in a reasonable period of time. If reaction is to occur at a mercury working electrode, then a mercury pool is used. For reaction at a solid surface, a platinum gauze electrode is generally employed.

If the polarographic or voltammetric wave is reversible, a potential on the plateau of the wave, 0·2 V after the half-wave potential is suitable for controlled potential electrolysis. If the wave is irreversible, complete oxidation or reduction will also result from controlled potential electrolysis at a potential 0·2 V beyond the half-wave potential, even though the plateau has not yet been reached. Of course, the time taken for the electrolysis will be lengthened if the chosen potential is on the rising part of an irreversible wave rather than on the plateau. If a polarogram or voltammogram consists of two well-separated waves, the second of which is totally irreversible, and if controlled potential electrolysis at a potential on the plateau of the first wave

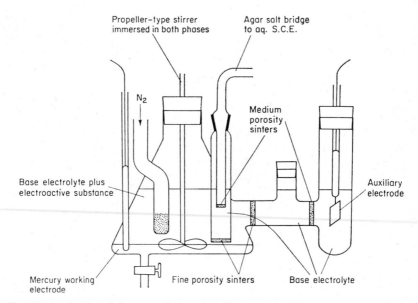

Fig. 23. A suitable cell for controlled potential electrolysis with non-aqueous solutions.

is required, care must be taken to see that the chosen potential is well-removed from the start of the second irreversible wave, for, at potentials at the start of such a wave, the oxidation or reduction is favoured thermodynamically and can proceed slowly even though the polarographic or voltammetric diffusion current for such a reaction appears to be negligible.

Information on solvents, base electrolytes, de-oxygenation of solutions, control of temperature, working ranges and reference electrodes is the same as that given in Chapter 3. However the currents passing through the solution can be quite large (hundreds of milliamps) and it is necessary to work with a three-electrode cell, the large current passing between the working and auxiliary electrodes, while the current passing between the working and reference electrode is negligible. A typical cell for non-aqueous solvents is shown in Fig. 23.

Similar cells can be designed for aqueous solutions or for a platinum gauze working electrode. The material of the auxiliary electrode should be chosen so that a suitable electrochemical reaction can proceed at this electrode during the electrolysis. With aqueous solutions, a platinum spiral immersed in the base electrolyte is convenient as an auxiliary electrode. If reduction is occurring at the working electrode, it is advisable to add hydrazine to the compartment containing a platinum auxiliary electrode to prevent oxidation of the platinum, as the hydrazine is then oxidized instead.

$$N_2H_4 \rightarrow N_2 + 4H^+ + 4e.$$

In non-aqueous solvents when reduction is occurring at the working electrode, a coil of copper can be used as the auxiliary electrode.

Controlled potential electrolysis is used when the product of the electrochemical reaction cannot be readily produced by a conventional oxidizing or reducing agent added to a solution of the starting material, or if it is desirable not to contaminate the solution with the reaction products of such an agent. The solution of the oxidation or reduction product can be examined by ultraviolet, visible and e.s.r. spectroscopy, polarography, and so on.

When the total quantity of electricity flowing during the electrolysis is determined, the technique is called controlled potential coulometry and a current integrator is inserted in series with the cell. Alternatively the current may be passed through a standard resistance in series with the cell and the potential drop across the resistance determined on a potentiometric recorder. This produces a graph of current against time and the area under the curve is integrated to give the total quantity of electricity which has passed during the electrolysis. To maintain a smooth fall in current during the electrolysis, the stirrer should be attached to a constant speed motor.

The value of the initial current is

$$i_0 = \frac{nFDAC}{\delta} \qquad (6.1)$$

where i_0 = initial current (mA); D = diffusion coefficient of the electroactive species (cm^2 sec^{-1}); A = electrode area (cm^2); C = initial concentration (mmoles l^{-1}); and δ = thickness of the diffusion layer (cm).

During the electrolysis the current falls exponentially according to the equation

$$i = i_0 \exp(-\beta t) \qquad (6.2)$$

where i is the current after time t and β equals $DA/V\delta$, V being the volume of the solution. A plot of log i versus t is, therefore, a straight line of slope $-\beta/2 \cdot 303$ and intercept log i_0.

The total quantity of electricity which has passed during the electrolysis, is given by

$$Q = \int_0^\infty i \, . \, dt. \qquad (6.3)$$

The electrolysis is actually stopped when i equals $0.001 \, i_0$, since at this stage the oxidation or reduction is virtually complete. Then

$$n = \frac{Q}{Fm} \qquad (6.4)$$

where Q is the quantity of electricity in coulombs, F is the Faraday (96,490 C) and m is the amount of material taken in moles. Of course, controlled potential electrolysis is also an important technique for quantitative analysis. Under these circumstances Q is measured, F and n are known, and the amount of electroactive material is calculated.

If n values are determined only occasionally, measuring the area under the current–time curve is perfectly adequate, but when n values have to be determined for a large number of compounds, it is convenient to use a current –time integrator in series with the cell. These can be electronic or electromechanical in design and a number of such instruments are commercially available; they are described fully by Lingane [21] and Rechnitz.[22] However, there is always much to be said for recording a current-time curve for each electroactive substance because, if the current does not fall in the way to be expected from Eqn (6.2) after the first few seconds of electrolysis, then the system is not so straightforward as had been thought and n values should not be computed from the total quantity of electricity resulting from the electrolysis.

In the most exact work, the quantity of electricity required to oxidize or reduce trace impurities in the base electrolyte and to charge the electrical double layer at the surface of the electrode should also be determined, and a correction for this blank applied. Alternatively the trace impurities may be oxidized or reduced in the base electrolyte in the cell, before adding and dissolving a known amount of the electroactive substance in the base electrolyte.

It must be appreciated of course that Eqn (6.2) does not apply if the oxidized or reduced species is regenerated by reaction of the product with the solvent or base electrolyte, or if the base electrolyte or solvent is slowly oxidized or reduced at the potential being applied to the working electrode.

The potential of the working electrode with respect to the reference electrode is selected before the electrolysis commences and is maintained constant by means of a potentiostat. The author uses the potentiostat, which is part of the Beckman Electroscan 30, but many commercial potentiostats are available or a potentiostat can be built in the laboratory. For further details on potentiostats the books by Lingane [21] and Rechnitz,[22] and the article by Lott [97] should be consulted. However, it is as well to remember that, with potentiostats, the maximum output voltages operating between the working and auxiliary electrodes, and the current capacities, vary widely from one instrument to another. Preferably the instrument should be capable of producing the current stipulated by Eqn (6.1). When this current is not being produced by the potentiostat, the graph of current against time starts as a straight line parallel to the time axis instead of falling exponentially to the axis. When this occurs it is usually the output voltage of the potentiostat that is insufficient to produce the current set by Eqn (6.1). This is most likely to happen with non-aqueous solvents where the resistance between the working and auxiliary electrodes can be appreciable. If, for example, this resistance is 1000 Ω and the maximum voltage which can be supplied by the potentiostat is 10 V, then the current cannot exceed 10 mA, even though Eqn (6.1) may stipulate a current greatly in excess of this value. To overcome this problem, an instrument with a greater output voltage should be used.

It must also be stressed that, although reversible polarograms and cyclic voltammograms are often obtained for an electroactive species, it cannot be assumed that the product of the electron transfer reaction is stable indefinitely. This product may react slowly with its surroundings after diffusing away from the working electrode, and this possibility must be borne in mind. Of course, this is most likely to occur with products which are powerful reducing or oxidizing agents, than with products which are produced at a working electrode within the potential range of $+0.5$ to -1.0 V *versus* the aq. S.C.E. Information on the stability of solvents to such substances is given on pages 63 and 67.

A few examples of the ways in which controlled potential electrolysis and coulometry have been used by inorganic chemists will serve to illustrate the usefulness of this technique. Rechnitz [98] has prepared aqueous solutions of the tetrachlororuthenium (II) anion by controlled potential reduction of the hexachlororuthenium (IV) anion in 4 M potassium chloride solutions adjusted to pH 1.5 with hydrochloric acid, at a mercury pool cathode at -0.47 V *versus* aq. S.C.E. Controlled potential coulometry gave an *n* value of 2.01 ± 0.01. Dessy and co-workers [80, 81] have used controlled potential electrolysis and coulometry widely in their work with organometallic compounds of the transition elements to prepare solutions of oxidized and reduced species, and to determine *n* values. Controlled potential electrolysis has also been used to prepare solutions of paramagnetic transition metal complexes for e.s.r. measurements.[99, 100, 101]

For further information on controlled potential electrolysis and coulometry, the reader is referred to the books and articles by Meites,[16, 96] Lingane,[21] Rechnitz [22] and Kies.[102]

CHAPTER 7

Aqueous Solutions

MORE electrochemical measurements have been made in aqueous solution than in all other solvents taken together, and numerous compilations of formal and standard electrode potentials for couples in water have been published.[10, 103, 104, 105] These contain information for simple inorganic cations, anions and molecules, and for a few complex ions. The formal electrode potentials for many other couples, of which at least one of the two forms is a complex ion, can be extracted from tables of polarographic data.[19, 106] For example, extensive data are available for the polarographic behaviour of inorganic species in base electrolytes such as citrate, cyanide, ethylenediamine, oxalate and pyridine. However, it must be remembered that these data have been accumulated by analytical chemists, who are primarily interested in polarography as a technique for quantitative analysis. The analytical chemist strives to keep the system as simple as possible consistent with his requirements, and the complexed species, with which he works, are not necessarily those of immediate interest to the modern inorganic chemist working with organometallic compounds, and so on. No extensive collection of polarographic data for organometallic compounds, dithiolene complexes, and so on, has been made and it is the desire of the author to remedy this situation.

If the inorganic chemist has synthesized a new compound and wishes to study its redox behaviour electrochemically, he should then consider water as a solvent in the first instance, since it is more convenient to use than other solvents. Of course, the compound must be soluble in water to give a saturated solution of concentration preferably $\geqslant 10^{-3}$ M, and it must not react with water. The potential ranges which are covered by polarography and voltammetry in various base electrolytes are shown in Table 7.

With the dropping mercury electrode, the range is terminated at the positive end by the oxidation of mercury and at the negative end by the reduction of hydrogen ion at pHs less than about two, and by the reduction of the cation of the base electrolyte at pHs greater than about two. In the pH range 2–6, hydrogen ion is reduced irreversibly with a half-wave potential of $-1\cdot56$ V *versus* the aq. S.C.E. The reduction of water is very irreversible on mercury.

With the rotating platinum electrode the range is terminated at the positive end by the oxidation of water or the anion of the base electrolyte, for example, bromide. At the negative end, the range is usually terminated by the reduction of hydrogen ion or water. The reduction of water proceeds much more readily on platinum than on mercury.

TABLE 7

Polarographic and Voltammetric Potential Ranges in Various Aqueous Base Electrolytes

Electrode	Base electrolyte	Potential range *versus* aq. S.C.E. (V).
D.M.E.	0·1 M $NaClO_4$ or 0·1 M $NaNO_3$	$+0·4$ to $-1·9$
D.M.E.	0·1 M NaCl	0 to $-1·9$
D.M.E.	0·1 M Me_4NClO_4	$+0·4$ to $-2·8$
D.M.E.	0·1 M Me_4NCl	0 to $-2·8$
D.M.E.	0·1 M HCl	0 to $-1·2$
D.M.E.	10 M HCl	0 to $-0·9$
D.M.E.	0·1 M NaOH	$-0·1$ to $-1·9$
R.P.E.	0·1 M $NaClO_4$, 0·1 M NaCl or 0·1 M Me_4NClO_4	$+1·0$ to $-1·0$
R.P.E.	0·1 M $HClO_4$	$+1·2$ to $-0·3$
R.P.E.	0·1 M NaOH	$+0·7$ to $-1·2$

For many purposes, these potential ranges seem to be very satisfactory and polarographic waves can be obtained for inorganic species in neutral or basic solutions of tetramethylammonium salts at potentials as negative as $-2·7$ V *versus* aq. S.C.E. However, the information obtained from polarographic waves at these negative potentials is often less useful than it appears to be at first sight. At very negative potentials, the waves may be partly catalytic in character, since the reduced species can react rapidly with water to regenerate the oxidized species. Also having obtained a formal electrode potential, the inorganic chemist will wish to put his information to good use. For example he may wish to prepare the reduced species in appreciable concentrations for further study by using a suitable reductant such as, say, sodium amalgam, or by controlled potential electrolysis. However, the reduced species cannot be prepared in aqueous solution, if it is oxidized by water at an appreciable rate.

The region of thermodynamic stability of inorganic species in aqueous solution is shown in Fig. 24. Since the oxidation and reduction of water is kinetically hindered to some extent, this region can be extended upwards and downwards for about 0·5 V on the potential axis, but an inorganic species forming half of a reversible couple will react with water, if the formal electrode potential for the couple falls outside this extended region. For these reasons it is better to study the redox behaviour of substances, which are reduced at fairly negative potentials, in solvents such as propylene carbonate which are more stable to reduction than water.

To summarize, therefore, the inorganic chemist will find it convenient to use aqueous solutions for electrochemical work, if the species under study is soluble in water, and if the species and its oxidized or reduced form do not react with water. If these conditions do not hold, then a non-aqueous solvent should be used. In certain cases, it is convenient to use an aqueous-organic

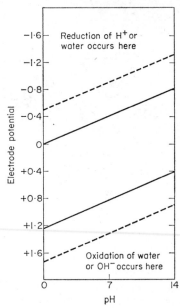

FIG. 24. Region of stability of oxidizing and reducing agents in aqueous solutions. (Reprinted with permission from Latimer, "Oxidation Potentials", 1952, Prentice-Hall, New Jersey.)[7]

solvent, because the compound under investigation is insoluble in water alone but, of course, these solvent mixtures should again only be employed when the compound and its oxidized or reduced form are non-reactive with water.

Polarographic, voltammetric and potentiometric data for a number of inorganic compounds of interest to the inorganic chemist are shown in Tables 8 and 9 for aqueous solutions and aqueous-organic mixtures respectively.

Some of the couples in Table 8 are models for biological systems. For example, the formal electrode potentials of copper (I,II) proteins and iron (II,III) porphyrin proteins, which are related to iron (II,III) dipyridyl complexes, have been reported by Brill et al.[121] and James et al.[122] respectively. Further information on the redox behaviour of metalloporphyrins is given by Falk and Phillips.[251]

TABLE 8

Polarographic, Voltammetric and Potentiometric Data for Aqueous Solutions

Redox system	Base electrolyte	Reference electrode	E^o (V versus R.E.)	References
$[(\pi\text{-}C_5H_5)_2Fe]^+ \to [complex]^0$	0.1 M $NaClO_4$	S.C.E.	+0.16	107
$[(\pi\text{-}C_5H_5)_2Ru]^+ \to [complex]^0$	0.1 M $NaClO_4$	S.C.E.	+0.11	107
$[(\pi\text{-}C_5H_5)_2Co]^+ \to [complex]^0$	0.1 M $NaClO_4$	S.C.E.	-1.16	107
$[(\pi\text{-}C_5H_5)_2Rh]^+ \to [complex]^0$	0.1 M $NaClO_4$	S.C.E.	-1.53	108
$[(\pi\text{-}C_5H_5)_2Ni]^+ \to [complex]^0$	0.1 M $NaClO_4$	S.C.E.	-0.21	109
$[(\pi\text{-}C_5H_5)_2Ti\,OH]^+ + H^+ + e \leftrightarrow \frac{1}{2}[(\pi\text{-}C_5H_5)_2Ti]^+ + H_2O$	0.1 M $NaClO_4$ / -0.1 M $HClO_4$	S.C.E.	-0.44	110, 111
$[(\pi\text{-}C_5H_5)_2V]^{2+} \leftrightarrow [complex]^+$	1 M HCl	S.C.E.	-0.32	111
$[(\pi\text{-}indenyl)_2Co]^+ \to [complex]^0$	0.1 M $NaClO_4$	S.C.E.	-0.6	112
$[(\pi\text{-}C_6H_6)_2Cr]^+ \to [complex]^0$ [a]	0.5 M LiCl	S.C.E.	-0.97	113
$[(\pi\text{-}toluene)_2Cr]^+ \to [complex]^0$ [a]	0.5 M LiCl	S.C.E.	-1.04	113
$[Fe(1,10\text{-}phen)_3]^{2+} \leftrightarrow [complex]^{3+}$ [b]	0.1 M KNO_3	N.H.E.	+1.13 [d]	114
	1 M H_2SO_4	N.H.E.	+1.06 [d]	115
$[Fe(5,6\text{-}Me_2\text{-}1,10\text{-}phen)_3]^{2+} \leftrightarrow [complex]^{3+}$	1 M H_2SO_4	N.H.E.	+0.97 [d]	116
$[Fe(5,6\text{-}Me_2\text{-}1,10\text{-}phen)_3]^{2+} \leftrightarrow [complex]^{3+}$	1 M H_2SO_4	N.H.E.	+1.25 [d]	115
$[Fe(4,7\text{-}Me_2\text{-}1,10\text{-}phen)_3]^{2+} \leftrightarrow [complex]^{3+}$ [c]	0.1 M $LiClO_4$	S.C.E.	+0.83	117, 118
$[Fe(2,2'\text{-}dipy)_3]^{2+} \leftrightarrow [complex]^{3+}$ [b]	0.1 M KNO_3	N.H.E.	+1.10 [d]	114
$[Ru(2,2'\text{-}dipy)_3]^{3+} \leftrightarrow [complex]^{3+}$ [b]	0.5 M H_2SO_4	N.H.E.	+1.25 [d]	119
$[Os(2,2'\text{-}dipy)_3]^{3+} \leftrightarrow [complex]^{3+}$ [b]	0.1 M KNO_3	N.H.E.	+0.88 [d]	114
$[Os(1,10\text{-}phen)_3]^{2+} \leftrightarrow [complex]^{3+}$ [b]	0.1 M KNO_3	N.H.E.	+0.89 [d]	114

[a] Half-wave potential data for other arene chromium compounds in 0.5 M aqueous lithium chloride have also been reported.[113, 120] [b] Many other formal electrode potentials for similar systems in water, methanol, acetonitrile and formamide have been reported.[114] [c] The formal electrode potential of this couple has also been determined in many organic solvents.[118] [d] These formal electrode potentials were determined by potentiometry; otherwise by polarography or voltammetry. A single arrow (→) denotes that polarographic or voltammetric data were obtained only with solutions of the species in front of the arrow. A double arrow (↔) denotes that the polarographic or voltammetric behaviour of both the oxidized and reduced species were studied separately, or that the formal electrode potential was determined potentiometrically.

TABLE 9

Polarographic and Potentiometric Data for Inorganic Species in Aqueous-Organic Solvents

Redox system	Base electrolyte	Reference electrode	$E^{o'}$ (V versus R.E.)	References
$[(\pi\text{-}C_5H_5)_2Fe]^0 \leftrightarrow [\text{complex}]^+$	90% EtOH, 0·1 M NaClO$_4$-0·01 M HClO$_4$	aq. S.C.E.	+0·31	107
$[(\pi\text{-}C_5H_5)_2Ni]^0 \rightarrow [\text{complex}]^+$	90% EtOH, 0·1 M NaClO$_4$	aq. S.C.E.	-0·08	109, 110
$[(\pi\text{-}C_5H_5)_2Os]^0 \leftrightarrow [\text{complex}]^+$	50% EtOH, 0·5 M HClO$_4$	aq. S.C.E.	-0·07	123
$[(\pi\text{-}C_5H_3)_2Co]^+ \leftrightarrow [\text{complex}]^0$	50% EtOH, 0·5 M HClO$_4$	aq. S.C.E.	-0·98	124
$[(\pi\text{-}C_5H_5)_2Fe]^0 \leftrightarrow [\text{complex}]^+$	75% HOAc, 0·034 M HClO$_4$	aq. S.C.E.	+0·23 c	37
^{a}Substituted ferrocenes $\leftrightarrow$ substituted ferricinium ions:				
p-methoxyphenyl-phenyl	75% HOAc, 0·034 M HClO$_4$	aq. S.C.E.	+0·23 c	37
p-bromophenyl-	75% HOAc, 0·034 M HClO$_4$	aq. S.C.E.	+0·26 c	37
p-chlorophenyl-	75% HOAc, 0·034 M HClO$_4$	aq. S.C.E.	+0·30 c	37
p-acetylphenyl	75% HOAc, 0·034 M HClO$_4$	aq. S.C.E.	+0·30 c	37
p-nitrophenyl-	75% HOAc, 0·034 M HClO$_4$	aq. S.C.E.	+0·33 c	37
	75% HOAc, 0·034 M HClO$_4$	aq. S.C.E.	+0·38 c	37
π-(3)-1,2-dicarbollyl metal derivatives:				
$[(\pi\text{-}B_9C_2H_{11})_2Fe]^- \rightarrow [\text{complex}]^{2-}$	50% acetone, 0·1 M LiClO$_4$	aq. S.C.E.	-0·42	6
$[(\pi\text{-}B_9C_2H_9Me_2)_2Fe]^- \rightarrow [\text{complex}]^{2-}$	50% acetone, 0·1 M LiClO$_4$	aq. S.C.E.	-0·54	6
$[(\pi\text{-}B_9C_2H_{10}Ph)_2Fe]^- \rightarrow [\text{complex}]^{2-}$	50% acetone, 0·1 M LiClO$_4$	aq. S.C.E.	-0·46	6
$[(\pi\text{-}B_9C_2H_{10}Ph)_2Co]^- \rightarrow [\text{complex}]^{2-}$	50% acetone, 0·1 M LiClO$_4$	aq. S.C.E.	-1·28	6
^{b}[Fe(oxine)$_3$]$^0 \leftrightarrow$[complex]$^-$	50% dioxan, 0·3 M NaCl	N.H.E.	-0·25 c	125

^{a} The formal electrode potentials of many other couples formed from substituted ferrocenes and their corresponding ferricinium ions have been measured in aqueous-organic solvents.[123, 126, 127] ^{b} Formal electrode potentials for couples formed from many substituted 8-hydroxy-quinoline complexes of iron (II) and (III) have also been reported.[125] ^{c} Formal electrode potentials determined potentiometrically; otherwise by polarography. The significance of single ($\rightarrow$) and double ($\leftrightarrow$) arrows is explained at the bottom of Table 8.

CHAPTER 8

Non-Aqueous Solutions

Choosing the Correct Solvent

As mentioned in Chapter 7, electrochemical investigations in aqueous solution are seldom undertaken by the inorganic chemist outside the range of +1·0 to −1·0 V *versus* the aq. S.C.E. If the reduced form of a couple with formal electrode potential less than −1·0 V *versus* the aq. S.C.E. is to be stable in solution, then a solvent other than water must be used. Some solvents which are more stable to reduction than water are acetonitrile, 1,2-dimethoxy-ethane, dimethylformamide, propylene carbonate and sulpholane. Less is known about solvents which are more stable to oxidation than water, but hydrogen fluoride and dichloromethane are certainly in this category. The polarographic and voltammetric ranges obtained with many non-aqueous solvents are shown in Table 10.

Although some of these ranges are extensive, it cannot be assumed that the reduced form of a couple with a very negative formal electrode potential will be stable in these solvents with even more negative reduction decomposition potentials, or that the oxidized form of a couple with a very positive formal electrode potential will be stable in these solvents with even more positive oxidation decomposition potentials. Without doubt the polarographic or voltammetric oxidations or reductions of most of the solvents in Table 10 are kinetically hindered to varying degrees. Some information about the long-term stabilities of some of these solvents to powerful oxidants and reductants is available.

The formal electrode potential of the *tris*-2,2′-dipyridyl iron (II)–*tris*-2,2′-dipyridyl iron (III) couple in acetonitrile, 0·1 M in lithium perchlorate, is approximately +0·9 V *versus* the aq. S.C.E.[114, 118] The blue-green iron (III) complex is stable in acetonitrile;[141] however the iron (III) complex is unstable in propylene carbonate [141] and oxidizes the solvent, being itself reduced to the red iron (II) complex. Therefore the oxidized form of a reversible couple is unstable in propylene carbonate if the electrode potential of the couple exceeds about +0·9 V *versus* aq. S.C.E. even though the working range extends to +2·2 V *versus* aq. S.C.E. in 0·1 M lithium perchlorate at a rotating platinum electrode (see Table 10). The stability of acetonitrile to moderately

TABLE 10

Polarographic and Voltammetric Ranges for Non-aqueous Solvents

| Solvent | Range (V versus aq. S.C.E.) and base electrolyte | | References |
	D.M.E.	R.P.E.	
Dimethylsulphoxide	+0·3 to −2·8(0·1 M Et$_4$NClO$_4$)	+0·7 to −1·9(0·1 M NaClO$_4$)	128
Dimethylformamide	approx +0·4 to −2·8(0·1 M Et$_4$NClO$_4$)	approx +1·7 to −1·9(0·1 M NaClO$_4$)	129
Benzonitrile	+0·5 to −2·1(0·1 M Et$_4$NClO$_4$)	—	130
Acetic acid-acetic anhydride (20 : 1, V/V)	approx +0·6 to −1·0(0·25 M NaClO$_4$)	—	30
Acetonitrile	+0·6 to −2·8(0·1 M Et$_4$NClO$_4$)	+1·8 to −1·5(0·1 M NaClO$_4$)	131, 132
1 : 2-Dimethoxyethane	approx +0·6 to −2·9(0·1 M Bu$_4$NClO$_4$)	—	133
Propylene carbonate	+0·6 to −2·9(0·1 M Et$_4$NClO$_4$)	+2·2 to −3·2(0·1 M LiClO$_4$)	134
Sulpholane	approx +0·6 to −2·9(0·1 M Et$_4$NClO$_4$)	approx +2·7 to −1·9(0·1 M NaClO$_4$)	135
Acetone	+0·7 to −2·5(0·1 M Et$_4$NClO$_4$)	—	136
Nitromethane	+0·7 to −1·3(0·1 M Et$_4$NClO$_4$)	+2·2 to −2·6[0·1 M Mg(ClO$_4$)$_2$]	134, 137
Nitrobenzene	+0·8 to −1·2(0·1 M Et$_4$NClO$_4$)	—	134
Dichloromethane	+0·8 to −1·3(0·05 M Et$_4$NClO$_4$)	+2·0 to −1·8(0·1 M Pr$_4$NClO$_4$)	134, 138
Pyridine	−0·6 to −2·2(0·15 M Bu$_4$NI)	+1·5 to −0·4(0·1 M LiClO$_4$)[a]	139, 140

[a] Pyrolitic graphite electrode. Where approximate values are given, the ranges have been determined against some other reference electrode and approximately corrected to the aq. S.C.E. using thallium (I) as pilot ion (see page 71).

powerful oxidants is verified by the fact that a solution of copper (II) in acetonitrile is stable.[142] The formal electrode potential for the copper (II)–copper (I) couple in acetonitrile is approximately $+1\cdot0$ V *versus* aq. S.C.E. in $0\cdot1$ M tetraethylammonium perchlorate.[143]

Some of the solvents in Table 10 are stable to quite powerful reducing agents. For example, the biphenyl radical anion [biphenyl]$^-$ is quite stable in 1,2-dimethoxyethane.[144] The formal electrode potential of the [biphenyl]$^{0/-1}$ couple in this solvent, which is $0\cdot1$ M in tetrabutylammonium perchlorate is $-3\cdot3$ V *versus* a reference electrode of silver/silver ion in the same solvent, that is about $-2\cdot6$ V *versus* the aq. S.C.E.[81] When one considers that the ytterbium (II) ion is rapidly oxidized by water, the formal electrode potential for the ytterbium (II)–ytterbium (III) couple being $-1\cdot4$ V *versus* the aq. S.C.E., it is obvious that a switch from water to a solvent such as 1,2-dimethoxyethane greatly increases the possibilities of preparing solutions of really powerful reducing agents.

Of course, it must be appreciated that considerable care is needed if a solvent is to be kept free of water when using electrochemical techniques. For inert complexes, when the formal electrode potentials of couples lie within the range of $+1\cdot0$ to $-0\cdot7$ V *versus* the aq. S.C.E., traces of water in the solvent are not likely to be of much consequence. However, for couples outside this range, traces of water are important and should be eliminated, if at all possible. A 10^{-3} M solution of the reduced form of a reversible couple of formal electrode potential of $-1\cdot5$ V *versus* the aq. S.C.E. will often be stable in anhydrous 1,2-dimethoxyethane, but if the solvent is made 10^{-2} M in water, the species will be rapidly oxidized by water.

Shrivington[145] has stated that it is not difficult to obtain acetonitrile $<2\times10^{-3}$ M in water but, after the anhydrous base electrolyte has been added and "dry" oxygen-free nitrogen passed through a solution for several minutes, the water content of a solution is usually in excess of 10^{-2} M. This shows that one must be particularly careful when determining n values by controlled potential coulometry, if the electrolytic product is a powerful oxidant or reductant; a trace of water in the solvent can lead to the wrong conclusions. This also applies to hydrolysable materials. Of course, the electrochemical behaviour of compounds which are hydrolysed by water can be studied in suitable non-aqueous solvents, but the possible presence of trace amounts of water in the non-aqueous solvents must be borne in mind.

Another important factor in choosing the correct solvent is that of solubility. The material under investigation must be soluble to the extent of at least 10^{-4} M; also the concentration of base electrolyte must be at least $0\cdot05$ M. Tetraalkylammonium perchlorates are suitable as base electrolytes in all solvents, and sodium perchlorate, which is much cheaper, can be employed in solvents of reasonable coordinating ability. The coordinating

ability of solvents for cations decreases in the following order, dimethylsulphoxide > dimethylformamide ⇌ water > acetone ⇌ propylene carbonate
⇌ acetonitrile > sulpholane > nitromethane > nitrobenzene > dichloromethane. Sodium perchlorate can be used as a base electrolyte in sulpholane and
solvents of greater coordinating ability.

The solvent must also have a useful liquid range and preferably its dielectric constant must not be too low as this results in very high cell resistance.
The physical properties of useful non-aqueous solvents are shown in Table 11.

TABLE 11

Physical Properties of Non-aqueous Solvents

Solvent	Liquid range (°C)	Dielectric constant	Viscosity (cp)
Acetonitrile	−45 to 82	38(20°)	0·38(15°)
Dichloromethane	−97 to 40	9 (20°)	0·39 (30°)
1,2-Dimethoxyethane (glyme)	−58 to 82	3·5	—
Dimethylformamide	−61 to 153	38 (25°)	0·80 (25°)
Dimethylsulphoxide	18 to 189	47 (25°)	1·99 (25°)
Nitromethane	−29 to 101	36 (30°)	0·62 (25°)
Propylene carbonate	−49 to 242	64 (25°)	2·53 (25°)
Sulpholane (tetrahydrothiophen 1,1-dioxide)	28 to 285	44 (30°)	9·87 (30°)

Finally another factor which must be considered with certain species is the
coordinating ability of the solvent. For example, a particular inorganic
complex may be unstable in a strongly coordinating solvent such as dimethylsulphoxide. A less basic solvent must therefore be employed for electrochemical investigations of this species; in this respect, dichloromethane is
particularly suitable.

Methods of purification to produce anhydrous solvents suitable for electrochemical studies have been described for acetonitrile,[146, 147] dimethylsulphoxide,[128] sulpholane,[135] dimethylformamide,[129, 147] 1,2-dimethoxyethane [133]
and dichloromethane.[148] To obtain propylene carbonate of suitable quality,
allow it to stand over sodium metal for two days shaking occasionally; then
decant the solvent from the metal and distil *in vacuo*, the middle cut of 90%
being used.[134] To purify nitromethane, cool the solvent in a mixture of solid
carbon dioxide and acetone until about 75% of the liquid solidifies. Reject
the liquid and repeat the fractional freezing twice. Stand the purified
solvent over phosphorus pentoxide for 24 hr and distil *in vacuo*; use the middle
cut of 75%.[134]

Much useful information on non-aqueous solvents used in electrochemical
studies is also given by Charlot and co-authors.[149] A good review on

techniques with non-aqueous solvents has been written by Popov,[150] and an article on electrochemistry in dimethylsulphoxide has been published recently.[151]

The Comparison of Electrode Potentials in Different Solvents

When formal electrode potentials are determined in aqueous solutions with the saturated calomel electrode as reference electrode, the correction for liquid junction potential between the aqueous solution and saturated potassium chloride solution is usually only a few millivolts [152] and can generally be neglected. However the liquid junction potentials between, say, 0·1 M aqueous sodium perchlorate and non-aqueous solvents are often quite appreciable and cannot be overlooked. There is no direct means of measuring these liquid junction potentials and, if a formal electrode potential determined for a couple in, say, acetonitrile is to be expressed *versus* the potential of the aqueous saturated calomel electrode without liquid junction potential, then some means of estimating the liquid junction potential must be found. They are usually estimated using a pilot ion method, although a graphical method developed by Vlček [153] has certain advantages over the pilot ion method and should receive further study.

Formal electrode potentials for a couple in various solvents *versus* the aq. S.C.E. without liquid junction potential are of considerable interest because by comparing these potentials much information can be obtained about the free energy changes associated with the transfer of an ion from one solvent to another. This will be discussed later.

The pilot ion method

The pilot ion to be used in this method should be one for which the free energy of solution of the gaseous ion † at 25°C is identical in all solvents plus base electrolytes; obviously it is impossible to find such an ion but large singly-charged cations best fit the purpose. Such ions are the rubidium, caesium, thallium (I), ferricinium and cobalticinium ions; the first three ions are reduced reversibly at a mercury cathode to their respective metal amalgams. If the free energy of solution of gaseous rubidium ions at 25°C is assumed to be identical in all non-aqueous solutions, then the absolute formal electrode potential of the Rb^+/Rb (Hg) couple will be invariable, as will be the potential of this couple *versus* the potential of any reference electrode without liquid junction potential. Similar arguments apply to caesium and thallium (I) ions. In the case of the ferricinium ion, it is reduced reversibly to ferrocene.

---

† In this book the free energy of solvation of the gaseous ion refers to the free energy change, which occurs when the gaseous ion is dissolved in a solvent to give a solution of zero ionic strength. When the gaseous ion is dissolved in a base electrolyte of appreciable ionic strength, the free energy change is referred to as the free energy of solution of the gaseous ion.

If the free energy of solution of the gaseous neutral molecules is identical in all solvents, and if the free energy of solution of the gaseous ions is also invariable in all solvents, then again the formal electrode potential of the $[(\pi\text{-}C_5H_5)_2Fe]^{+1/0}$ couple will be invariable. Similar arguments apply to the $[(\pi\text{-}C_5H_5)_2Co]^{+1/0}$ couple.

The formal (and standard) electrode potentials of these M^+/M^0 couples do change slightly in different solvents, [114, 154, 155] but they are much more nearly constant than the formal electrode potentials of M^{3+}/M^0, M^{2+}/M^0, M^{3+}/M^{2+}, M^{3+}/M^+ and M^{2+}/M^+ couples.

The values of some liquid junction potentials estimated by the pilot ion method for aqueous/non-aqueous solutions are shown in Table 12.

If measurements of formal electrode potentials have been made for couples in a non-aqueous solvent *versus* the aq. S.C.E., as is often the case,

TABLE 12

Liquid Junction Potentials for Aqueous/Non-aqueous Solutions Estimated by the Pilot Ion Method

Liquid junction		Magnitude of *l.j.p.* with the pilot ion or the species producing the pilot ion in parenthesis. $E^{0\prime}$ (non-aq. solution)$-E^{0\prime}$ (aq. solution)	References
Non-aq. solution	Aq. solution		
Dimethylformamide			
0·1 M in TEAP	0·1 M KNO_3	0 (Tl^+)	134
Acetic acid			
1 M in $LiClO_4$	sat. KCl	0 (Tl^+)	156
N-Methylacetamide			
0·1 M in TEAP	sat. KCl	+0·04 (Tl^+)	157
Dimethylsulphoxide			
0·1 M in TEAP	sat. KCl	+0·05 (K^+)	128, 158
Propylene carbonate			
0·1 M in TEAP	0·1 M $NaClO_4$	+0·08 (Tl^+), +0·09 (ferrocene)	134
Nitromethane			
0·1 M in TEAP	0·1 M $NaClO_4$	+0·13 (Tl^+), +0·05 (ferrocene)	134
Acetone			
0·1 M in TEAP	sat. KCl	+0·16 (Rb^+)	136
Nitrobenzene			
0·1 M in TEAP	0·1 M $NaClO_4$	+0·18 (Tl^+), +0·14 (ferrocene)	134
Acetonitrile			
0·1 M in TEAP	sat. KCl	+0·19 (Tl^+), +0·15 (Rb^+)	5
Dichloromethane			
0·05 M in TEAP	0·1 M $NaClO_4$	+0·20 (ferrocene)	134

TEAP is tetraethylammonium perchlorate. The formal electrode potentials were determined polarographically for the reduction of thallous and rubidium ions and for the oxidation of ferrocene.

then a correction for the liquid junction potential is readily applied to express these potentials *versus* the aq. S.C.E. without liquid junction potential. For example, the formal electrode potential of the $Cd^{2+}/Cd(Hg)$ couple in propylene carbonate, 0.1 M in sodium perchlorate, is -0.31 V *versus* the aq. S.C.E. Using the thallous ion as pilot ion, the liquid junction potential is $+0.08$ V, that is $E^{0\prime}_{P.C.} - E^{0\prime}_{H_2O}$ for the $Tl^+/Tl(Hg)$ couple. Hence $E^{0\prime}$ for $Cd^{2+}/Cd(Hg)$ in propylene carbonate, 0.1 M in sodium perchlorate, is -0.39 V *versus* the aq. S.C.E. without liquid junction potential.

When measurements of formal electrode potentials have been made for couples in a non-aqueous solvent *versus* a reference electrode in the same solvent, it is also necessary to correct for the liquid junction potential between the aqueous S.C.E. and the non-aqueous solution, and for the potential between the reference electrode in the non-aqueous solvent and the aqueous S.C.E., if these potentials are to be expressed *versus* the aq. S.C.E. without liquid junction potential. The following example will illustrate this point.

The formal electrode potential of the $Pb^{2+}/Pb(Hg)$ couple in dimethylformamide, 0.1 M in sodium perchlorate, is -0.03 V *versus* the $Ag/AgCl/0.1$ M tetraethylammonium chloride in dimethylformamide reference electrode. Its potential *versus* the aq. S.C.E. without liquid junction potential is required. The formal electrode potential of the $Tl^+/Tl(Hg)$ couple in dimethylformamide, 0.1 M in sodium perchlorate, is -0.07 V *versus* the Ag/AgCl/TEACl reference electrode. Subtracting 0.39 V from this potential gives the potential of the $Tl^+/Tl(Hg)$ couple in dimethylformamide *versus* the aq. S.C.E. without liquid junction potential, that is, -0.46 V, which is assumed to be independent of the solvent. Subtracting 0.39 V from the above potential of the $Pb^{2+}/Pb(Hg)$ couple gives -0.42 V, which is the potential of this couple in dimethylformamide, 0.1 M in sodium perchlorate, *versus* the aq. S.C.E. without liquid junction potential. The correction factor is made up of two parts, (1) the liquid junction potential between saturated potassium chloride in water and 0.1 M sodium perchlorate in dimethylformamide, and (2) the difference in potential between the aq. S.C.E. and the Ag/AgCl/TEACl reference electrode in dimethylformamide.

Vlček's graphical method[153]

This method is described for the determination of the liquid junction potential between water and glacial acetic acid.[159] A cell with three compartments is used.

+Hg	(1) Hg₂Cl₂, sat. KCl in H₂O	(2) HOAc + LiCl in H₂O	(3) sat. LiCl, Hg₂Cl₂ in HOAc	Hg⁻
		π_1	π_2	

π_1 and π_2 are the liquid junction potentials. The ratio [LiCl] : [HOAc] = 1 : 6·87 in compartment (3). This ratio is also maintained in compartment (2), but the lithium chloride and acetic acid concentrations are increased in steps from dilute solutions to about 5 M in acetic acid and the cell potential is measured.

For solutions up to 5 M in acetic acid in compartment (2), π_1 is negligible. When [HOAc] in compartment (2) reaches 17·25 M (glacial acetic acid), π_2 has disappeared. Now it has been found that up to 5 M acetic acid in compartment (2), cell potential plotted against log [HOAc] produces a straight line, and this line is extrapolated to [HOAc] = 17·25 M. This gives the cell potential in the absence of liquid junction potential, that is the electrode potential of the acetic acid calomel electrode *versus* the aq. S.C.E. without liquid junction potential. Other formal electrode potentials referred to the acetic acid calomel electrode can, therefore, be expressed against the aq. S.C.E. without liquid junction potential.

Electrochemical Behaviour of Simple Cations

Half-wave potentials for the reduction of some simple cations in water and in non-aqueous solvents are shown in Table 13 *versus* the aq. S.C.E. without liquid junction potential. For these waves, $E_{\frac{1}{4}} - E_{\frac{3}{4}}$ is less than or equal to $68/n$ mV, so that the waves are assumed to be reversible or almost so. Some could be quasireversible but this is unlikely since, in all cases where the oxidations of the metal amalgams have been investigated using dropping amalgam electrodes, $E_{\frac{1}{2}}$ (reduction) equalled $E_{\frac{1}{2}}$ (oxidation) for the appropriate couples. These values can therefore be taken as formal electrode potentials. The half-wave potentials were corrected for liquid junction potentials using the pilot ion method with thallous ion.

When these formal electrode potentials obtained in different solvents are compared, it is at once apparent that the potential of a couple usually becomes more positive as the coordinating ability of the solvent for the cation decreases. This is expected from theory (see below). The approximate order of coordinating ability of solvents for cations is dimethylsulphoxide > dimethylformamide > water > acetone > propylenecarbonate > acetonitrile > nitromethane > dichloromethane as has been ascertained from the heat of reaction of these solvents with acceptor molecules.[162, 163, 164] If the values for couples in Table 13 are compared the following order of potentials with sign reversed is obtained, dimethylsulphoxide > dimethylformamide ≃ water > propylene carbonate ≃ acetonitrile > acetic acid > sulpholane > nitromethane. Where applicable this closely follows the order of coordinating ability given above.

It must be stressed however that the formal electrode potentials in Table 13 are a measure of the free energy of solution of the gaseous ions in these

TABLE 13

Formal Electrode Potentials for Some Simple Ions in Water and in Non-aqueous Solvents

Solvent	Couples and formal electrode potentials (V $versus$ aq. S.C.E.)								References
	$Tl^{I/0}$	$Pb^{II/0}$	$Cd^{II/0}$	$Zn^{II/0}$	$Mn^{II/0}$	$Ba^{II/0}$	$Yb^{III/II}$	$Sm^{III/II}$	
Dimethylsulphoxide	−0·46	−0·47	−0·62	−0·98		−1·99			158
a Water	−0·46	−0·38	−0·58	−1·00	−1·46	−1·94	−1·41	−1·80	—
b Dimethylformamide	−0·46	−0·42	−0·56	−0·98	−1·51	−2·01	—	−1·84	—
c Acetonitrile	−0·46		−0·45	−0·81	−1·27	−1·82	−0·76	−1·17	—
Propylene carbonate	−0·46	−0·31	−0·40			−1·90	−1·07		134
Acetic acid	−0·46	−0·29	−0·36						156
Sulpholane	−0·46	−0·27	−0·27		−1·16				135
Nitromethane	−0·46	−0·24	−0·25						134

a Half-wave potential data in non-complexing base electrolytes from Meites,[160] except for manganese in 0·1 M sodium perchlorate solution.[160a] b Half-wave potential data from reference 129, except $Sm^{III/II}$ from reference 161. c Half-wave potential data from Coetzee et al.[5] except for $Ba^{II/0}$ from reference 161. Except dimethylsulphoxide (base electrolyte 0·1 M sodium nitrate for thallium (I), lead (II), cadmium (II) and zinc (II)), water (note a) and acetic acid (base electrolyte 1 M lithium perchlorate), the solutions were 0·1 M in perchlorate which is non-complexing, except perhaps in solvents of very low coordinating ability.

solvents plus base electrolytes. Unlike standard electrode potentials, they do not reflect directly changes in the free energies of solvation of the cations. This is evident from a study of the thermodynamic cycle in Fig. 25.

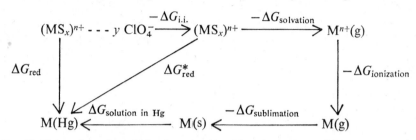

FIG. 25. Thermodynamic cycle for the reduction of ions in a solvent. $\Delta G_{i.i.}$ = free energy change resulting from ionic interactions; $\Delta G_{solvation}$ = free energy of solvation of the gaseous cations; $\Delta G_{ionization}$ = free energy change associated with the ionization of the gaseous atoms; $\Delta G_{sublimation}$ = free energy of sublimation; $\Delta G_{solution\ in\ Hg}$ = free energy change associated with dissolving the metal in mercury; ΔG_{red} = free energy change associated with the reduction at a mercury cathode (finite ionic strength); ΔG^*_{red} = free energy change associated with the reduction at a mercury cathode (zero ionic strength).

The coordinating ability of a solvent for cations is best expressed by the free energy of solvation of the gaseous ions at 25°C. For a cation in two solvents A and B

$$\Delta G^*_{red}(A) - \Delta G^*_{red}(B) = [-\Delta G_{solvation}(A)] - [-\Delta G_{solvation}(B)].$$

But $\Delta G^*_{red} = -n\,FE^*$ where E^* is the absolute standard electrode potential. Hence

$$-nF[E^*(A) - E^*(B)] = [-\Delta G_{solvation}(A)] - [-\Delta G_{solvation}(B)].$$

If the coordinating ability of A is greater than B then

$$-\Delta G_{solvation}(A) > -\Delta G_{solvation}(B).$$

Hence $E^*(B) - E^*(A)$ is positive or $E^*(B) > E^*(A)$.

The absolute standard electrode potential of the couple $(MS_x)^{n+}/M(Hg)$ in solvent B is therefore more positive than in solvent A. The standard electrode potential of the couple *versus* any reference electrode is therefore more positive in solution B (less coordinating) than in solution A (more coordinating). As the coordinating ability of a solvent for cations decreases, that is as the free energy of solvation of the gaseous ions becomes less negative, the standard electrode potential of the M^{n+}/M^0 couple becomes more positive. Using similar thermodynamic cycles it can be shown that the standard electrode potentials for $M^{n+}/M^{(n-a)+}$ couples also become more positive as the coordinating ability of the solvent decreases ($a = 1, \ldots, n-1$).

It is the formal electrode potentials which are measured polarographically, and these become more positive as the free energies of solution of gaseous ions become less negative. The free energy of solution of gaseous ions is $\Delta G_{ionization} + \Delta G_{i.i.}$.

Frequently, but by no means always, a decrease in the coordinating ability of a solvent, that is a decrease in $-\Delta G_{solvation}$, is paralleled by a decrease in its dielectric constant and an increase in $-\Delta G_{i.i.}$. Hence there are two opposing free energy terms here. Usually the change in $\Delta G_{i.i.}$ on going from a more coordinating to a less coordinating solvent is considerably less than the change in $\Delta G_{solvation}$, and so the values of formal electrode potentials for couples, formed from at least one cation, become, like the values of standard electrode potentials, more positive as the coordinating ability of the solvent decreases.

The effect of the $\Delta G_{i.i.}$ term has, in the past, been frequently neglected and for many systems this is not justified. For aqueous solutions, $\Delta G_{i.i.}$ is small for non-complexing anions such as perchlorate (a few electron centivolts at the most) but it cannot be neglected for solutions of base electrolytes in solvents of much lower dielectric constant such as acetic acid ($\varepsilon = 6$).

Very few data are available on the differences between standard and formal electrode potentials for couples in non-aqueous solvents. However, for the tris-1,10-phenanthroline iron (III)/tris-1,10-phenanthroline iron (II) couple in acetonitrile, Kolthoff and Thomas [165] have found a difference of 0·08 V between the standard and formal (0·1 M perchlorate) electrode potentials, and these are large cations. For smaller doubly charged and triply charged cations the effects of ionic interactions will be appreciable in solvents of low dielectric constant.

Finally, it is important to remember that, by the pilot ion method, liquid junction potentials are only estimated and not determined exactly. The formal electrode potentials in Table 13 could well be in error by 100 mV, so that it is pointless to attach thermodynamic significance to differences of less than 100 mV in formal electrode potentials obtained in different solvents. However, when generous allowance is made for the shortcomings in the pilot ion method, there is no doubt that the formal electrode potentials of most cationic couples move to more positive potentials as the coordinating ability of the solvent for cations decreases.

Occasionally, however, the opposite happens. The formal electrode potentials of the $Cu^+/Cu(Hg)$ and $Ag^+/Ag(Hg)$ couples in acetonitrile, 0·1 M in tetraethylammonium perchlorate, are $-0·52$ V and $+0·23$ V versus the aq. S.C.E. without liquid junction potential, respectively (Tl^+ as pilot ion).[5] The standard electrode potentials of these couples in water are $+0·15$ V [166] and $> +0·4$ V [160] versus the aq. S.C.E. respectively. Acetonitrile complexes copper (I) and silver (I) ions more strongly than water does. These are termed specific effects, and they are rather uncommon.

The inorganic chemist is, of course, interested in putting these shifts in formal electrode potential to good use, possibly by synthesizing in solution ions of unusual oxidation state. Solvents, which coordinate more strongly than water, are complexing agents in aqueous solution and complexing agents have been widely used to stabilize higher oxidation states in aqueous solution, for example the stabilization of silver (II) as the tetrapyridinesilver (II) ion. The use of non-aqueous solvents is, therefore, not always necessary for preparing ions in these unusual higher oxidation states; working in aqueous solutions with the complexing agent is often possible.

If the formal electrode potential for a couple in the presence of the complexing agent is more negative than $+1.0$ V *versus* the aq. S.C.E., then the oxidized form of the couple is likely to be stable in aqueous solution. However, if the formal electrode potential for a couple in the absence of a complexing agent is more positive than $+2.0$ V *versus* the aq. S.C.E., it is seldom possible to stabilize the oxidized form of the couple by adding a complexing agent sufficiently for it not to react with water, because few complexing agents are able to shift the formal electrode potential of a couple by more than one volt in a negative direction.

It should therefore be possible to stabilize some simple solvated ions of interesting higher oxidation states in non-aqueous solvents which coordinate cations more strongly than water and which themselves are more stable than water to oxidation. Species such as copper (III) and americium (IV) come to mind. Finding solvents which coordinate cations more strongly than water is no problem, but finding solvents that do this and are also stable to powerful oxidants, is more difficult. Liquid hydrogen fluoride is in this class.

Cations of unusual lower oxidation states are stabilized in solvents for which the free energy of solution of gaseous ions is less negative than in aqueous solution. For example, the formal electrode potentials of the samarium (III)/samarium (II) and samarium (II)/samarium amalgam couples in 0.1 M aqueous tetramethylammonium iodide are -1.80 V and about -1.96 V *versus* the aq. S.C.E. respectively,[160] and in acetonitrile, 0.1 M in tetraethylammonium perchlorate, are -1.17 V and about -1.88 V *versus* the aq. S.C.E. without liquid junction potential, respectively (Tl^+ as pilot ion).[5] The samarium (II) state has been appreciably stabilized to oxidation in acetonitrile.

It is interesting to speculate on the possibility of preparing in acetonitrile, or in other solvents of low coordinating ability, cations which disproportionate in aqueous solution. The two ions that immediately come to mind are copper (I) and indium (I) because reliable information on the formal electrode potentials of the couples involving these ions in aqueous solution is available. These are

Cu^{2+}/Cu^+ $E^0 = -0.08$ V *versus* the S.C.E.[166]

$Cu^+/Cu(Hg)$ $E_a^0 = +0.15$ V *versus* the S.C.E.[166]

In^{3+}/In^+ $E^{0\prime} = -0.66$ V *versus* the S.C.E. (3 M $NaClO_4$) [167]

$In^+/In(Hg)$ $E_a^{0\prime} = -0.15$ V *versus* the S.C.E. (3 M $NaClO_4$) [167, 168]

Very approximately, the free energy of solution of gaseous cations, excluding the hydrogen ion, in a particular solvent is proportional to the square of the charge on the ion. This being so, the differences in the formal electrode potentials of couples in two solvents of different coordinating abilities decrease in the order $M^{3+}/M^{2+} > M^{3+}/M^{+} > M^{3+}/M^{0} \rightleftharpoons M^{2+}/M^{+} > M^{2+}/M^{0} > /M^{+}/M^{0}$, provided that there are no specific effects. In Table 13 the differences in the $E^{0\prime}$ values for M^{3+}/M^{2+} couples in water and acetonitrile are certainly much greater than the differences in $E^{0\prime}$ values for M^{2+}/M^{0} couples.

In a solvent such as nitromethane, one would expect the formal electrode potential for the $In^{+}/In(Hg)$ couple to be slightly more positive than -0.15 V *versus* the aq. S.C.E. without liquid junction potential, but the formal electrode potential of the In^{3+}/In^{+} couple to be much more positive than -0.66 V. If the shifts in potentials in nitromethane are such that $E^{0\prime}$ (In^{3+}/In^{+}) $- E^{0\prime}$ (In^{+}/In (Hg)) is now greater than 0.2 V, then In (I) will be almost completely stable to disproportionation. In fact, Ashraf and Headridge [134] have prepared solutions of indium (I) in nitromethane by oxidizing indium amalgam to indium (I) with a deficiency of anhydrous silver perchlorate. From such solutions, white crystals of indium (I) perchlorate are readily prepared by pumping off the solvent.

In acetone, 0.1 M in tetraethylammonium perchlorate, the formal electrode potentials of the Cu^{2+}/Cu^{+} and $Cu^{+}/Cu(Hg)$ couples are $+0.37$ V and $+0.15$ V *versus* the aq. S.C.E. without liquid junction potential respectively (Rb^{+} as pilot ion).[136] Copper (I) ion is obviously quite stable to disproportionation in acetone.

It is queried whether other cations of lower oxidation state, hitherto unknown in solution, can be prepared in suitable non-aqueous solvents. Among the rare earths, solid compounds of europium (II), ytterbium (II), samarium (II), thulium (II) and neodymium (II) are known. The polarographic reduction of the foremost three in aqueous solution produces first the divalent ion and then the amalgam. All three amalgams, ytterbium (II) and samarium (II) are rapidly oxidized by water and cannot be isolated in aqueous solution. Polarographic reduction of thulium (III) and neodymium (III) in aqueous solution shows no evidence for the formation, even momentarily, of thulium (II) or neodymium (II). Since the positive shift in the formal electrode potential of a III/II couple is much greater than that for a II/O(Hg) couple on transfer of the ions from water to a solvent of lower coordinating ability, the polarographic reduction of thulium (III) and neodymium (III) in solvents of lower coordinating ability than water, might be expected to produce evidence for thulium (II) and neodymium (II) in such solvents. However in acetonitrile [5, 130] and benzonitrile [130] both thulium (III) and neodymium (III) produce but one three-electron reduction wave showing that thulium (II) and neodymium (II) will, in the presence of mercury, disproportionate to the trivalent ion and the amalgam even in these solvents.

However, the zero oxidation state of rare earth metals, like the alkali metals, is appreciably stabilized to oxidation by amalgam formation, and the disproportionation of thulium (II) and neodymium (II) in solution will be aided if mercury is added to the system. The best chance of producing these species in solution will be by the controlled potential oxidation of the corresponding metal in a solvent, which shows relatively low coordinating ability for cations and which is stable to powerful reducing agents. Treating the rare earth metal in such a solvent with a deficiency of a suitable oxidizing agent could also produce the desired divalent ion. By similar techniques, gallium (I) and americium (II) may possibly be prepared in non-aqueous solvents of lower coordinating ability than water.

Complexing Agents in Non-Aqueous Solvents

In water with non-complexing base electrolytes such as 0·1 M sodium perchlorate, metal ions exist as hydrated complexes. With very few exceptions, these produce just one polarographic reduction wave when a one-stage electron transfer is involved. Solvated metal ions appear to reduce in a similar way in non-aqueous solvents, and for this reason metal perchlorates and perchlorate base electrolytes are frequently used in non-aqueous solvents. The perchlorate ion is only a weak complexing agent, if a complexing agent at all, in non-aqueous solvents useful for electrochemical studies. Hence the data given in the preceding section have usually been obtained in solutions where perchlorate is the only anion present. In non-aqueous solvents of lower coordinating ability than water, chloride and nitrate ions are much better complexing agents than in aqueous solution, and metal chlorides, nitrates, and so on, should not be employed when half-wave potentials are to be compared for different solvents. Solutions of the metal ions should then be prepared from perchlorates, tetrafluoroborates or hexafluorophosphates.

The half-wave potentials of reversible reduction waves for many metallic ions in non-aqueous solvents of lower coordinating ability than water, containing only perchlorate anions, are moved to more negative potentials on adding chloride, nitrate, and so on, as a result of complexation by these ions, for example the effect of nitrate ions on the polarographic reduction waves for lead and cadmium ions in molten dimethylsulphone.[169] Also, troublesome double or multiple polarographic waves can sometimes be obtained in non-aqueous solvents, when the reduction of a metal ion occurs in the presence of a stoichiometric amount of complexing agent or in the presence of a deficiency or slight excess of complexing agent. Two examples of such phenomena are as follows.

(1) For a 10^{-3} M solution of manganese (II) perchlorate in 0·1 M tetra-ethylammonium perchlorate in dimethylformamide, $E_{\frac{1}{2}} = -1·55$ V *versus* the saturated calomel electrode in DMF as reference electrode. On replacing some of the tetraethylammonium perchlorate with tetraethylammonium chloride, the first wave decreased in height and was gradually replaced by a second more irreversible wave at $E_{\frac{1}{2}} = -2·1$ V *versus* the reference electrode. For $[Cl^-]_{\text{total}}/[Mn^{(II)}]_{\text{total}} > 4$, the first wave disappeared indicating that the second wave was due to an irreversible reduction of $MnCl_4^{2-}$. With

$$0 < [Cl^-]_{\text{total}}/[Mn^{(II)}]_{\text{total}} < 4,$$

two waves were obtained.[170]

(2) For an approximately 10^{-3} M solution of anhydrous zinc nitrate in sulpholane, 0·1 M in sodium perchlorate, three polarographic waves were obtained with half-wave potentials of $-0·04$, $-0·18$ and $-0·35$ V *versus* a reference electrode of Ag/AgCl/saturated tetraethylammonium chloride in sulpholane. All the waves were very irreversible and the ratio of the wave heights was approximately $1 : 2 : 1$.[135] These waves have not been examined in detail by the methods outlined by Vlček,[12] but a mixture of complexes which reduce at different potentials may be present in the solution or the first waves may be kinetic in nature. Such difficulties do not occur in solutions free from complexing agents.

Of course, most of the complexes of interest to inorganic chemists are rather inert to dissociation in non-aqueous solvents and trouble from double waves of the type just mentioned does not occur. These inert complexes are considered in a later section.

Water is also a complexing agent in non-aqueous solvents which coordinate cations less readily than water. The stepwise stability constants of aquo-copper (II) complexes in nitromethane and acetone are given in Table 14.

TABLE 14

Stepwise Stability Constants of Aquo-copper (II) Complexes in Non-aqueous Solvents

Solvent	Base electrolyte	log k_1	log k_2	log k_3	log k_4	log k_5	log k_6	References
Nitromethane	0·1 M TEAP	2·90	1·92	1·22	0·94	0·52	0·80	171
Acetone	0·1 M TEAP	1·75	1·25	0·80	0·65	—	—	172

Electrochemical Behaviour of the Hydrogen Ion

The solvated hydrogen ion is reduced irreversibly on mercury in all solvents. However, by the direct potentiometric method, the standard electrode potentials for the hydrogen ion/hydrogen couple have been measured in several non-aqueous solvents and are expressed versus the aq. N.H.E.

without liquid junction potential in Table 15, by assuming that the standard electrode potential of the Rb$^+$/Rb couple is invariable at $-2 \cdot 93$ V *versus* the aq. N.H.E. in these solvents.[173]

TABLE 15

Standard Electrode Potentials of the Hydrogen Ion/Hydrogen Couple in Some Non-aqueous Solvents

Solvent	Water	Formic acid	Hydrazine	Methanol	Forma-mide	Aceto-nitrile
E_H^0 *versus* the aq. N.H.E.	0·00	+0·52	−0·92	−0·02	−0·08	+0·24

As expected, it will be noticed that the solvated hydrogen ion reduces most readily in solvents, which solvate it only weakly, that is formic acid and acetonitrile. However, in hydrazine, which solvates the hydrogen ion strongly, the reduction is less readily effected.

Although the hydrogen ion, whether solvated by the solvent or combined with an anion in an essentially undissociated acid, is reduced irreversibly at the dropping mercury electrode, the half-wave potential data for acids can be explained by considering the basicities of the solvent and the conjugate bases. For example in pyridine, which is a basic solvent, identical polarographic waves are obtained for acids whose pK_a values in aqueous solution are less than 7·9.[174] Acetonitrile, however, is a differentiating rather than a levelling solvent since it does not solvate the hydrogen ion strongly. Polarographic data for acids in acetonitrile are shown in Table 16. It will be noticed that the half-wave potentials of the waves move in a negative direction as the acid strength decreases. The polarographic reduction of hydrogen ion in non-aqueous solvents has been reviewed by Elving and Spritzer.[173]

TABLE 16

Polarography of Freshly Prepared 10^{-3} M Solutions of Acids in Acetonitrile [175]

Acid	$-E_{\frac{1}{2}}$ *versus* the aq. S.C.E. (V)
Perchloric	0·70
Hydrobromic	0·90
Hydrochloric	1·06
p-Toluenesulphonic	1·20
2,5-Diethylanilinium ion	1·43
Oxalic	1·55
Phosphoric	1·75
Benzoic	2·1
Acetic	2·3

Electrochemical Behaviour of Oxygenated Species

The simplest of these species is oxygen itself. In aqueous solution, oxygen is reduced in two steps, first to hydrogen peroxide and then to water.

$$O_2 + 2H^+ + 2e \rightarrow H_2O_2$$
$$H_2O_2 + 2H^+ + 2e \rightarrow 2H_2O.$$

However, the polarographic reduction products of oxygen in aprotic solvents were only established in 1965 and 1966. Oxygen is reduced in two stages in aprotic solvents. The first wave is reversible or almost so and occurs at about the same potential in aprotic solvents as shown in Table 17.[40, 176]

TABLE 17

Polarographic Data for the First Reduction Wave for Oxygen in Aprotic Solvents

Solvent	Base electrolyte	$-E_{\frac{1}{2}}$ (V versus aq. S.C.E.)
Dimethylsulphoxide	0·1 M Bu$_4$NClO$_4$	0·77
Dimethylformamide	0·1 M Bu$_4$NClO$_4$	0·87
Dichloromethane	0·2 M Bu$_4$NClO$_4$	0·79
Acetonitrile	0·2 M Bu$_4$NClO$_4$	0·82
Acetone	0·2 M Bu$_4$NClO$_4$	0·88
Pyridine	0·2 M Bu$_4$NClO$_4$	0·89

By controlled potential coulometry of oxygen in dimethylsulphoxide, 0·1 M in tetraethylammonium perchlorate, on the plateau of the first wave, it has been established that $n \simeq 1$, showing that the reduction product is the superoxide ion.[177] This has also been established by cyclic voltammetry and e.s.r. spectroscopy[176, 178]; and from the solution of superoxide ion produced by the controlled potential electrolysis of oxygen in dimethylformamide, 0·1 M in tetra-n-butylammonium perchlorate, on the plateau of the first wave, insoluble potassium superoxide has been isolated.[178]

At more negative potentials a second polarographic reduction wave for oxygen has been obtained in aprotic solvents, its half-wave potential being dependent on the cation of the base electrolyte. It has been postulated that the reduction product of the second wave is the peroxide ion. When metal ions are present in the solution, the peroxide is stabilized, that is more readily produced, as a result of its insolubility.[41]

$$O_2^- + 2M^+ + e \rightarrow M_2O_2.$$

More recently it has been shown using chronopotentiometry and controlled potential coulometry in conjunction with pH titrations, infrared spectroscopy and gas chromatography, that the peroxide ion first formed on the reduction of superoxide ion in dimethylsulphoxide in the presence of tetra-ethylammonium perchlorate as base electrolyte, reacts rapidly with the tetra-ethylammonium ion as follows[179]:

$$O_2^{2-} + (C_2H_5)_4N^+ \rightarrow HO_2^- + CH_2 {=} CH_2 + (C_2H_5)_3N$$

$$HO_2^- + DMSO \rightarrow DMSO_2 + OH^-.$$

The polarographic reductions of other oxyspecies in non-aqueous solutions are likely to be of interest to the inorganic chemist only if the reduction products are different from those in aqueous solution, or if a reduction product is a stable species, which disproportionates in aqueous solution or is oxidized by water. The polarographic behaviour of some oxyspecies, in aqueous and aprotic solvents, is shown in Table 18.

TABLE 18

A Comparison of the Polarographic Behaviour of Some Oxyspecies in Water and Non-aqueous Solvents

Species under investigation	UO_2^{2+}	VO^{2+}	ReO_4^-
Solvent	Water	Water	Water
Base electrolyte	0·5 M NaClO$_4$ −0·01 M HClO$_4$	2 M HOAc–2 M NH$_4$OAc 0·01 % (W/V) gelatin	2 M KCl
$E_{\frac{1}{2}}$ (V versus aq. S.C.E.)	−0·18	−1·24	−1·43
$E_{\frac{1}{4}} - E_{\frac{3}{4}}$ (mV)	56	60	—
Product and comments	UO_2^+ which disproportionates	V^{2+} $VO^{2+} + 2H^+ + 2e \rightarrow V^{2+} + H_2O$	H_2 Catalytic wave
References	19	19	180
Solvent	Dimethylsulphoxide	Acetonitrile	Acetonitrile
Base electrolyte	0·1 M TEAP	0·1 M TEAP	0·1 M TEAP
$E_{\frac{1}{2}}$ (V versus aq. S.C.E.)	−0·53	(1) −0·60 (2) −1·55	−2·58
$E_{\frac{1}{4}} - E_{\frac{3}{4}}$ (mV)	56	(1) 75 (2) 440	56
Product and comments	Probably UO_2^+ which may be stable	(1) Probably VO^+ (2) Probably VO	Probably ReO_4^{2-}
References	181	182	182

Further work on these species is desirable. Other oxygenated species which could well be investigated with the hope of producing new compounds are selenium dioxide and trioxide, tellurium dioxide and trioxide, sulphur trioxide, iodine pentoxide, chromium trioxide, chromate and dichromate ions, and so on.

Inert Complexes

In this monograph an inert complex is considered to be a complex which does not readily dissociate in solution. It is the opposite of a labile complex. The redox behaviour of such complexes is often simple (see Chapter 4) and formal electrode potentials can be obtained directly from potentiometric,

TABLE 19

Polarographic, Voltammetric and Chronopotentiometric Data for Metal-hydrocarbon Compounds in Non-aqueous Solutions

Redox system	Solvent and base electrolyte	Reference electrode	$E^{0\prime}$ (V versus R.E.)	References
a $[(\pi\text{-}C_6H_6)_2Cr]^+$ $\leftrightarrow[\text{complex}]^0$	80% MeOH–20% C_6H_6 0·3 M NaOH	0·5 M aq. calomel el.	$-0\cdot81$	183, 184, 185
a $[(\pi\text{-toluene})_2Cr]^+$ $\leftrightarrow[\text{complex}]^0$	80% MeOH–20% C_6H_6 0·3 M NaOH	0·5 M aq. calomel el.	$-0\cdot89$	184, 185
a $[(\pi\text{-mesitylene})_2Cr]^+$ $\leftrightarrow[\text{complex}]^0$	80% MeOH–20% C_6H_6 0·3 M NaOH	0·5 M aq. calomel el.	$-1\cdot00$	184, 185
$[(\pi\text{-}C_7H_7)Cr(\pi\text{-}C_5H_5)]^+$ $\leftrightarrow[\text{complex}]^0$	80% MeOH–20% C_6H_6 0·5 M NaOH	0·5 M aq. calomel el.	$-0\cdot70$	186
b $[(\pi\text{-}C_5H_5)_2Fe]^0$ $\rightarrow[\text{complex}]^+$	CH_3CN 0·1 M $LiClO_4$	aq. S.C.E.	$+0\cdot35$	118
$[(\pi\text{-}C_5H_5)_2Fe]^0$ $\rightarrow[\text{complex}]^+$	CH_3CN 0·2 M $LiClO_4$	aq. S.C.E.	$+0\cdot31$ d	85, 187
c Substituted ferrocenes $\rightarrow$Subst. ferricinium ions :				
1,1'-Et$_2$	CH_3CN 0·2 M $LiClO_4$	aq. S.C.E.	$+0\cdot19$ d	85, 187
-Et	CH_3CN 0·2 M $LiClO_4$	aq. S.C.E.	$+0\cdot25$ d	85, 187
-CH=CH$_2$	CH_3CN 0·2 M $LiClO_4$	aq. S.C.E.	$+0\cdot33$ d	85, 187
-COOH	CH_3CN 0·2 M $LiClO_4$	aq. S.C.E.	$+0\cdot55$ d	85, 187
-COMe	CH_3CN 0·2 M $LiClO_4$	aq. S.C.E.	$+0\cdot57$ d	85, 187
1,1'-(COMe)$_2$	CH_3CN 0·2 M $LiClO_4$	aq. S.C.E.	$+0\cdot80$ d	85, 187
$[(\pi\text{-}C_5H_5)_2Co]^+$ $\rightarrow[\text{complex}]^0$	Formamide 0·2 M $NaClO_4$	aq. S.C.E.	$-1\cdot11$	188
$[(\pi\text{-}C_6H_6)_2Cr]^+$ $\rightarrow[\text{complex}]^0$	Formamide 0·2 M $NaClO_4$	aq. S.C.E.	$-1\cdot04$	188
$[(\pi\text{-}C_6H_5.C_6H_5)_2Cr]^+$ $\rightarrow[\text{complex}]^0$	Formamide 0·2 M $NaClO_4$	aq. S.C.E.	$-0\cdot89$	188
$[(\pi\text{-}C_5H_5)_2TiCl_2]+e$ $\rightarrow[(\pi\text{-}C_5H_5)_2TiCl]+Cl^-$	DMF 0·1 M LiCl	aq. S.C.E.	$-0\cdot64$	189
$[(\pi\text{-}C_5H_5)$ V $(\pi\text{-}C_7H_7)]^0$ $\leftrightarrow[\text{complex}]^+$	CH_3CN 0·1 M TEAP	aq. S.C.E.	$+0\cdot19$	190

a Half-wave potential data for other arenechromium compounds in alcohol-benzene solutions have also been reported.[185, 191] b The formal electrode potential of this couple has been determined polarographically in many organic solvents.[118] c Quarter transition time potentials for other substituted ferrocenes have also been determined.[4, 192, 193, 194] d Formal electrode potentials determined by chronopotentiometry, otherwise by polarography or voltammetry. A single arrow ($\rightarrow$) denotes that polarographic, voltammetric or chronopotentiometric data were obtained only with solutions of the species in front of the arrow. A double arrow ($\leftrightarrow$) denotes that the polarographic behaviour of both the oxidized and reduces species were studied separately.

polarographic, voltammetric and chronopotentiometric data. There is considerable activity in this field at the present time and there is every reason to believe that interest in the electrochemistry of inert complexes is accelerating and will continue to do so in the future. Electrochemical data for many inert complexes in aqueous organic solutions have been given already in Chapter 7. Similar data for complexes in non-aqueous solutions are recorded in Tables 19, 20 and 22. These refer to metal-hydrocarbon compounds, dithiolene and related complexes and miscellaneous complexes, respectively.

TABLE 20

Polarographic and Voltammetric Data for the Reversible Waves of Metal-dithiolene and Related Complexes in Non-aqueous Solutions

Redox system	Solvent and base electrolyte	Reference electrode	$E^{0\prime}$ (V versus R.E.)	References
$[Ni(S_2C_2Ph_2)_2]^0$	DMSO			
$\leftrightarrow[complex]^-$	0·05 M	aq. S.C.E.	+0·22	195
$\rightarrow[complex]^{2-}$	n-Pr$_4$NClO$_4$		−0·74	195
a $\{Ni[S_2C_2(CN)_2]_2\}^{2-}$	CH$_3$CN			
$\rightarrow\{complex\}^-$	0·05 M n-Pr$_4$NClO$_4$	aq. S.C.E.	+0·23	195
b $\{Ni[S_2C_2(CF_3)_2]_2\}^0$	CH$_3$CN			
$\leftrightarrow\{complex\}^-$	0·05 M n-Pr$_4$NClO$_4$	aq. S.C.E.	+1·00	195
$\leftrightarrow\{complex\}^{2-}$			−0·12	195
c $\{Mo[S_2C_2(CF_3)_2]_3\}^0$	CH$_3$CN			
$\leftrightarrow\{complex\}^-$	0·05 M n-Pr$_4$NClO$_4$	aq. S.C.E.	+0·95	196
$\leftrightarrow\{complex\}^{2-}$			+0·36	196
d $\{Cr[S_2C_2(CN)_2]_3\}^{3-}$	CH$_3$CN			
$\leftrightarrow\{complex\}^{2-}$	0·05 M n-Pr$_4$NClO$_4$	aq. S.C.E.	+0·16	196
$\rightarrow\{complex\}^-$			+0·76	196
$[V(S_2C_2Ph_2)_3]^0$	CH$_3$CN			
$\leftrightarrow[complex]^-$	0·05 M n-Pr$_4$NClO$_4$	aq. S.C.E.	+0·30	197
$\rightarrow[complex]^{2-}$			−0·71	197
$\{V[S_2C_2(CF_3)_2]_3\}^{2-}$	CH$_3$CN			
$\leftrightarrow\{complex\}^-$	0·05 M n-Pr$_4$NClO$_4$	aq. S.C.E.	+0·47	197
$[V(C_6H_4S_2)_3]^{2-}$	CH$_3$CN			
$\rightarrow[complex]^{3-}$	0·05 M n-Pr$_4$NClO$_4$	aq. S.C.E.	−0·12	197
$[V(MeC_6H_3S_2)_3]^{2-}$	CH$_3$CN			
$\rightarrow[complex]^{3-}$	0·05 M n-Pr$_4$NClO$_4$	aq. S.C.E.	−0·18	197
e $[Pt(C_6H_4N_2H_2)_2]^0$	DMSO			
$\rightarrow[complex]^-$	0·1 M	aq. S.C.E.	−1·04	198
$\rightarrow[complex]^{2-}$	n-Pr$_4$NClO$_4$		−1·72	198
$\rightarrow[complex]^+$			+0·21	198
$\rightarrow[complex]^{2+}$			+0·77	198
$[Ni(C_6H_4NHS)_2]^0$				
$\rightarrow[complex]^-$	DMSO	aq. S.C.E.	−0·19	198
$\rightarrow[complex]^{2-}$	0·1 M n-Pr$_4$NClO$_4$		−1·04	198
$[Ni(C_6H_4OS)_2]^{2-}$	DMSO			
$\rightarrow[complex]^-$	0·1 M n-Pr$_4$NClO$_4$	aq. S.C.E.	−0·42	198

TABLE 20 (*contd.*)

Redox system	Solvent and base electrolyte	Reference electrode	$E^{0\prime}$ (V versus R.E.)	References
→[complex]^0			$+0.38$	198
[Ni(C$_6$H$_4$S$_2$)$_2$]$^-$	DMSO			
→[complex]$^{2-}$	0·1 M n-Pr$_4$NClO$_4$	aq. S.C.E.	-0.51	198
→[complex]^0			$+0.45$	198
^{f} [Cu(MeC$_6$H$_3$S$_2$)$_2$]$^-$	DMF	Ag/AgClO$_4$		
→[complex]$^{2-}$	0·1 M n-Pr$_4$NClO$_4$	0·1 M in DMF	-1.15	199
^{g} [Co(C$_6$H$_4$S$_2$)]$^-$	DMF	Ag/AgClO$_4$		
→[complex]$^{2-}$	0·1 M n-Pr$_4$NClO$_4$	0·1 M in DMF	-1.38	200
^{g} [Co(Me$_2$C$_6$H$_2$S$_2$)$_2$]$^-$	DMF	Ag/AgClO$_4$		
→[complex]$^{2-}$	0·1 M n-Pr$_4$NClO$_4$	0·1 M in DMF	-1.46	200
^{g} [Co(Me$_4$C$_6$S$_2$)$_2$]$^-$	DMF	Ag/AgClO$_4$		
→[complex]$^{2-}$	0·1 M n-Pr$_4$NClO$_4$	0·1 M in DMF	-1.57	200
^{g} [Co(C$_6$Cl$_4$S$_2$)$_2$]$^-$	DMF	Ag/AgClO$_4$		
→[complex]$^{2-}$	0·1 M n-Pr$_4$NClO$_4$	0·1 M in DMF	-0.85	200
[Re(S$_2$C$_2$Ph$_2$)$_3$]^0	DMF	Ag/AgClO$_4$		
→[complex]$^-$	0·1 M n-Pr$_4$NClO$_4$	0·1 M in DMF	-0.34	201
→[complex]$^{2-}$			-1.81	201
→[complex]$^{3-}$			-2.59	201
→[complex]$^+$			$+0.16$	201
[Re(MeC$_6$H$_3$S$_2$)$_3$]^0	DMF	Ag/AgClO$_4$		
→[complex]$^-$	0·1 M n-Pr$_4$NClO$_4$	0·1 M in DMF	-0.07	201
→[complex]$^{2-}$			-1.58	201
→[complex]$^{3-}$			-2.38	201
→[complex]$^+$			$+0.39$	201
^{h} [Pd(C$_6$H$_4$N$_2$H$_2$)$_2$]^0	DMSO			
→[complex]$^-$	0·05 M n-Pr$_4$NClO$_4$	aq. S.C.E.	-0.89	99
→[complex]$^{2-}$			-1.44	99
→[complex]$^+$			$+0.10$	99
→[complex]$^{2+}$			$+0.78$	99
[Ni(C$_6$H$_4$N$_2$H$_2$)$_2$]^0	Acetone			
→[complex]$^-$	0·05 M n-Pr$_4$NClO$_4$	aq. S.C.E.	-0.89	99
→[complex]$^{2-}$			-1.43	99
→[complex]$^+$			$+0.14$	99
[Ni(iPrC$_6$H$_3$N$_2$H$_2$)$_2$]^0	Acetone			
→[complex]$^-$	0·05 M n-Pr$_4$NClO$_4$	aq. S.C.E.	-0.98	99
→[complex]$^{2-}$			-1.51	99
→[complex]$^+$			$+0.09$	99
{Ni[(MeCNPh)$_2$]$_2$}$^{2+}$	CH$_3$CN			
→{complex}^{0}	0·05 M n-Pr$_4$NClO$_4$	aq. S.C.E.	-0.61	99
→{complex}$^-$			-1.60	99
→{complex}$^{2-}$			-1.80	99
^{i} {Co[S$_2$C$_2$(CF$_3$)$_2$]$_2$}$_2^0$	CH$_2$Cl$_2$	Ag/Ag I in		
→{Co[S$_2$C$_2$(CF$_3$)$_2$]$_2$}$_2^-$	0·1 M n-Bu$_4$NPF$_6$	CH$_2$Cl$_2$ 0·42 M	$+1.24$	202
→{Co[S$_2$C$_2$(CF$_3$)$_2$]$_2$}$_2^{2-}$		in n-Bu$_4$NPF$_6$	$+0.50$	202
or {Co[S$_2$C$_2$(CF$_3$)$_2$]$_2$}$^-$		and 0·05 M in		
→{Co[S$_2$C$_2$(CF$_3$)$_2$]$_2$}$^{2-}$		n-Bu$_4$NI	-0.20	202

TABLE 20 (contd.)

Redox system	Solvent and base electrolyte	Reference electrode	$E^{0\prime}$ (V versus R.E.)	References
i {Co[S$_2$C$_2$(CN)$_2$]$_2$}$^-$	CH$_2$Cl$_2$	Ag/Ag I in		
→{Co[S$_2$C$_2$(CN)$_2$]$_2$}$_2^-$	0·01 M n-Bu$_4$NPF$_6$	CH$_2$Cl$_2$ 0·42 M	+1·08	202
		n-Bu$_4$NPF$_6$ and		
		0·05 M in n-Bu$_4$NI		
[Ni(C$_6$H$_4$NHS)$_2$]^0	CH$_2$Cl$_2$	Ag/AgI in		
→[complex]$^-$	0·1 M n-Bu$_4$NPF$_6$	CH$_2$Cl$_2$ 0·5 M in	−0·03	100
→[complex]$^{2-}$		n-Bu$_4$NPF$_6$ and	−0·93	100
→[complex]$^+$		0·05 M in n-Bu$_4$NI	+1·05	100
j [Ni(dtbh)]^0	CH$_2$Cl$_2$	Ag/AgI in		
→[complex]$^-$	0·1 M n-Bu$_4$NPF$_6$	CH$_2$Cl$_2$ 0·5 M in	−0·48	100
→[complex]$^{2-}$		n-Bu$_4$NPF$_6$ and	−1·26	100
		0·05 M in n-Bu$_4$NI		
[Ni(SCPhNNH)$_2$]^0	DMSO			
→[complex]$^-$	0·05 M n-Pr$_4$NClO$_4$	aq. S.C.E.	−0·14	100
→[complex]$^{2-}$			−1·13	100
[Ni(dbh)]^0	DMSO			
→[complex]$^-$	0·05 M n-Pr$_4$NClO$_4$	aq. S.C.E.	−0·92	100
→[complex]$^{2-}$			−1·67	100
k [Ni(gma)]^0	DMSO			
→[complex]$^-$	0·05 M n-Pr$_4$NClO$_4$	aq. S.C.E.	−0·30	100
→[complex]$^{2-}$			−1·05	100
{Fe(NO)[S$_2$C$_2$(CF$_3$)$_2$]$_2$}$^-$	CH$_2$Cl$_2$	aq. calomel		
→{complex}$^{2-}$	0·1 M Et$_4$NClO$_4$	electrode	−0·07	101
→{complex}$^{3-}$		(0·1 M LiCl)	−0·36	101
→{complex}^{0}			+0·84	101
[Fe(NO)(C$_6$Cl$_4$S$_2$)$_2$]$^-$	CH$_2$Cl$_2$	aq. calomel		
→[complex]$^{2-}$	0·1 M Et$_4$NClO$_4$	electrode	−0·24	101
→[complex]^0		(0·1 M LiCl)	+0·74	101
[Fe(NO)(MeC$_6$H$_3$S$_2$)$_2$]$^-$	CH$_2$Cl$_2$	aq. calomel		
→[complex]$^{2-}$	0·1 M Et$_4$NClO$_4$	electrode	−0·64	101
→[complex]^0		(0·1 M LiCl)	+0·27	101
[Fe(NO)(S$_2$C$_2$Ph$_2$)$_2$]$^-$	CH$_2$Cl$_2$	aq. calomel		
→[complex]$^{2-}$	0·1 M Et$_4$NClO$_4$	electrode	−0·83	101
→[complex]^0		(0·1 M LiCl)	−0·02	101
→[complex]$^+$			+0·71	101
{Fe(NO)[S$_2$C$_2$(CN)$_2$]$_2$}$^{2-}$	CH$_2$Cl$_2$	aq. calomel		
→{complex}$^-$	0·1 M Et$_4$NClO$_4$	electrode	+0·03	101
		(0·1 M LiCl)		

a-k Data are reported for similar compounds of the following elements, a Co, Pt, Cu and Pd[195], b Pt and Pd[203], c Cr and W[196], d V[196], e Ni[198], f Ni, Pt, Co, Fe and Au[199], g Ni and Cu[200], h Pt and Co[99], i Fe[202], j Cu, Zn and Cd[100], and k Zn and Cd[100]. A single arrow (→) denotes that polarographic and voltammetric data were obtained only with solutions of the species in front of the arrow. A double arrow (↔) denotes that the polarographic and voltammetric behaviour of both the oxidized and reduced species were studied separately.

Key to Formulae in Table 20

$[M(S_2C_2Ph_2)_n]^0$

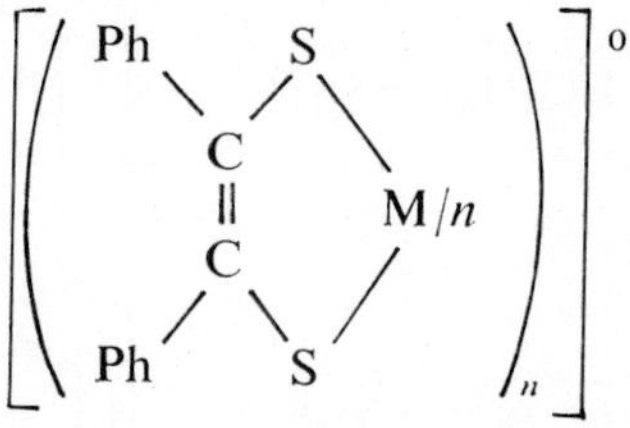

$\{M[S_2C_2(CN)_2]_n\}^{x-}$

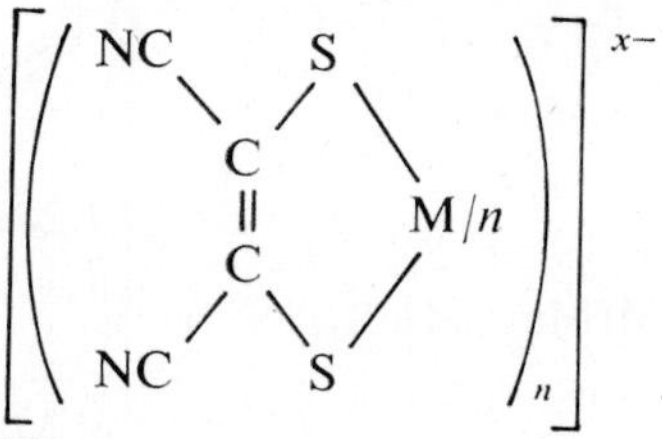

$\{M[S_2C_2(CF_3)_2]_n\}^0$

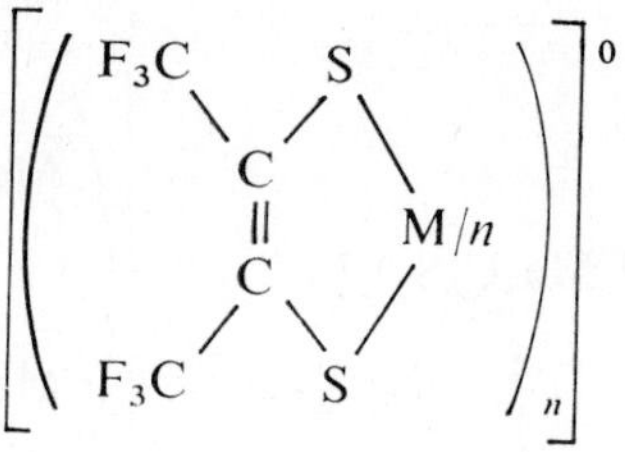

$[M(C_6H_4S_2)_n]^{x-}$

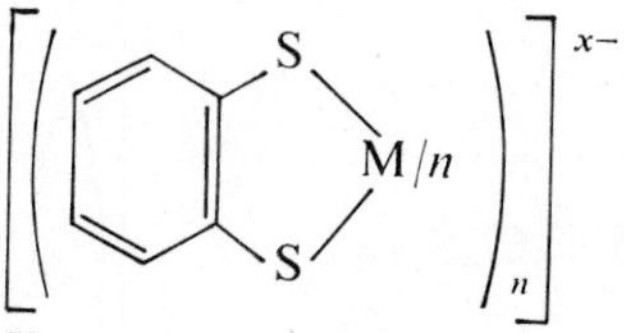

$[M(MeC_6H_3S_2)_n]^{x-}$

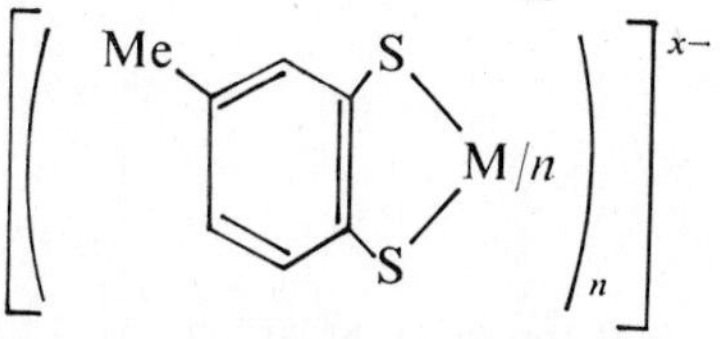

$[M(C_6H_4N_2H_2)_2]^0$

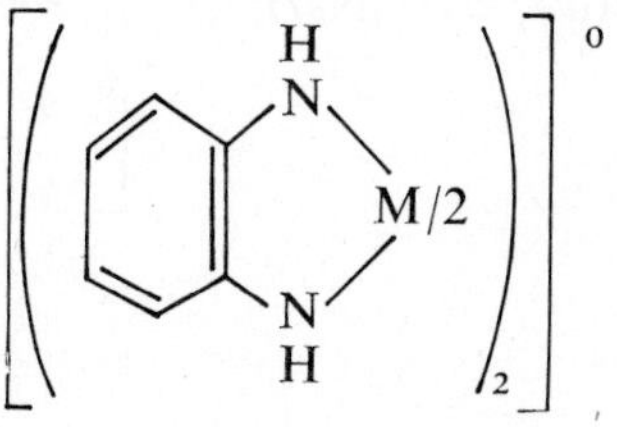

$[M(C_6H_4NHS)_2]^0$

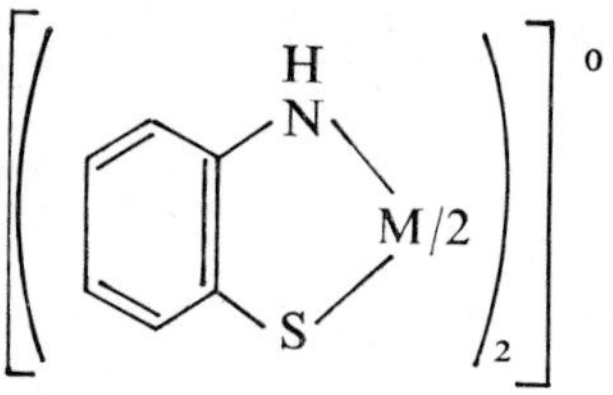

$[M(C_6H_4OS)_2]^{2-}$

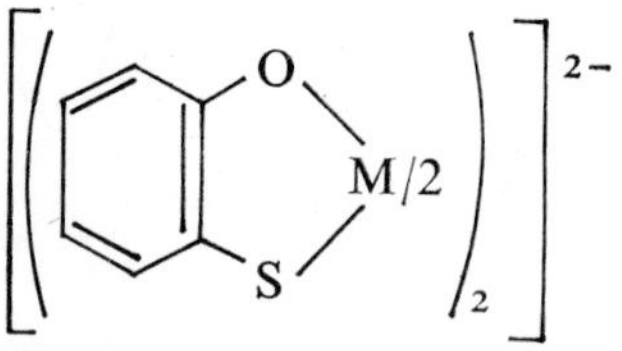

$[M(Me_2C_6H_2S_2)_2]^-$

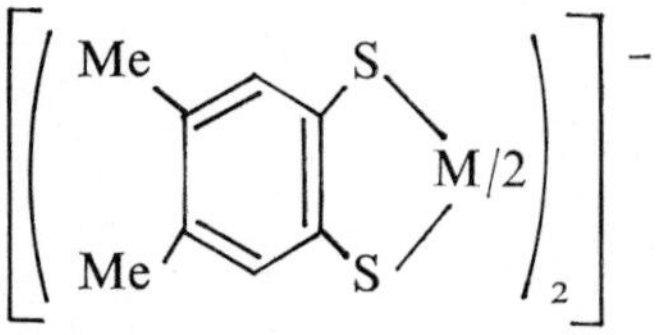

$[M(Me_4C_6S_2)_2]^-$

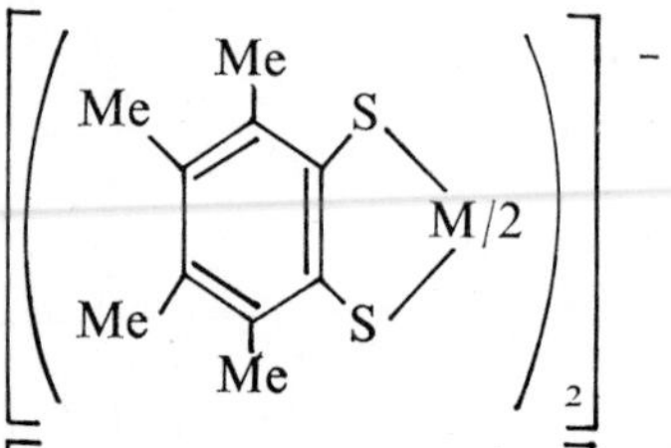

$[M(C_6Cl_4S_2)_2]^-$

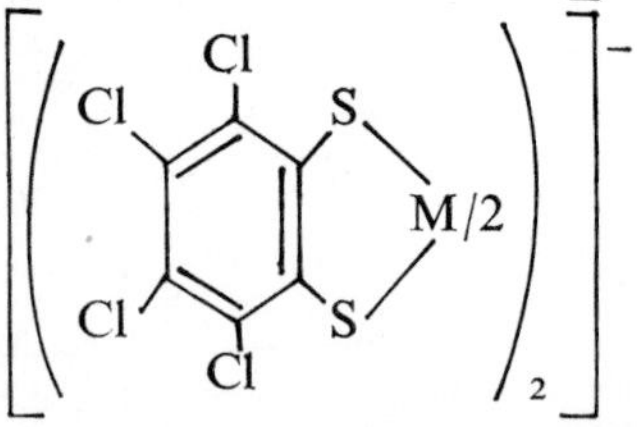

$[M(iPr.C_6H_3N_2H_2)_2]^0$

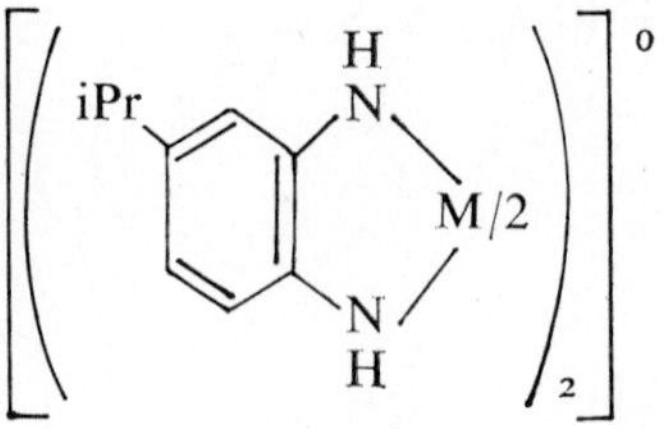

$\{M[(MeCNPh)_2]_2\}^{2+}$

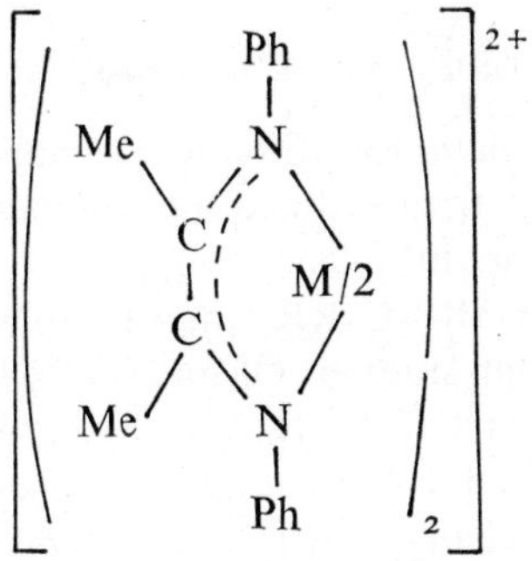

$[M(SCPhNNH)_2]^0$

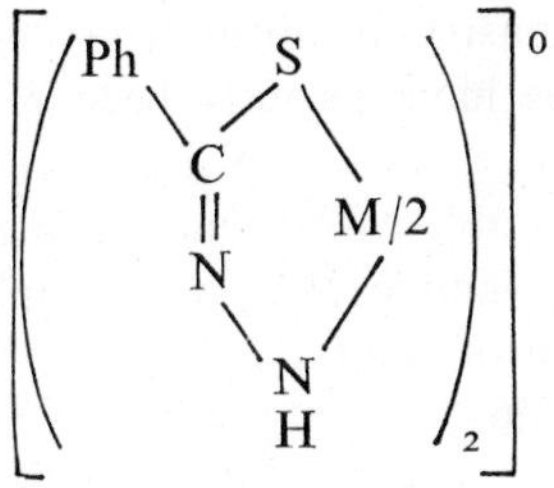

$[M(dtbh)]^0$

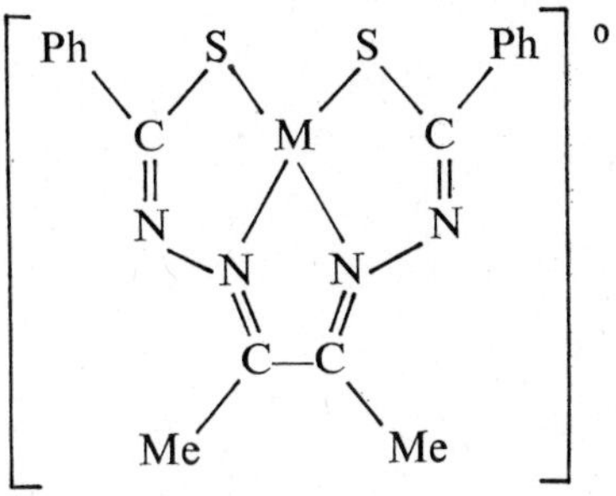

$[M(dbh)]^0$

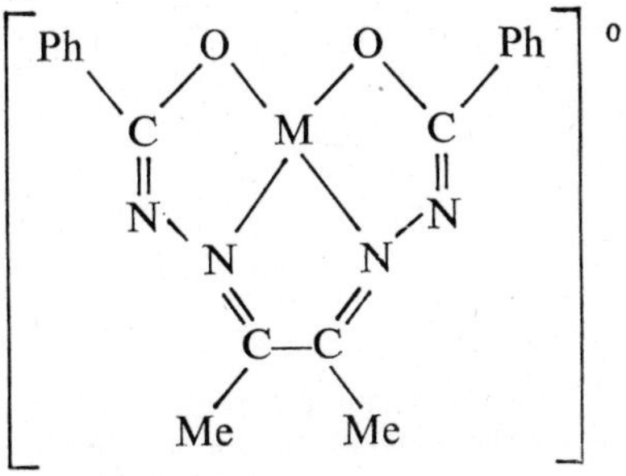

$[M(gma)]^0$

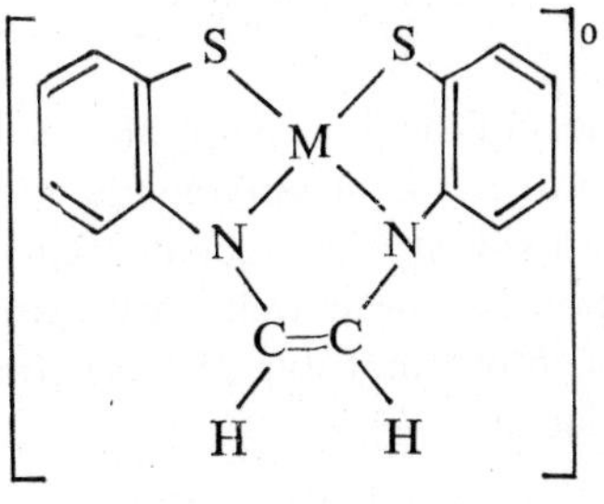

4*

Polarographic data for dithiolene complexes of iron and cobalt further coordinated by trisubstituted phosphines have been reported.[204] Polarographic data for dithiolene complexes of nickel, palladium and platinum, $M(S_2C_2R_2)_2$ or $M(S_2C_2RR')_2$ have also been obtained and related to structure in a thorough study by Olson *et al.*[3] where R and R' were CN, p-C_6H_4Cl, C_6H_5, H, p-$C_6H_4CH_3$, CH_3, p-$C_6H_4OCH_3$, C_2H_5, n-C_3H_7 and iso-C_3H_7. Polarography and voltammetry have also been used as tools to study the slow ligand-exchange reactions of *bis*-dithiolene nickel complexes.[205] Finally the chemistry of dithiolene complexes including their electron transfer properties have recently been reviewed in an excellent article by McCleverty.[206]

In certain favourable cases, much information can be obtained from one polarogram. For example, the polarographic waves arising from the oxidation and reduction of $\{Pd[C_6H_4(NH_2)_2]_2\}^0$ in dimethylsulphoxide solution [99] are shown in Fig. 26.

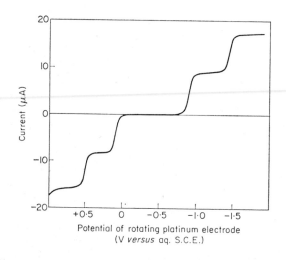

FIG. 26. A voltammogram for $\{Pd[C_6H_4(NH_2)_2]_2\}^0$ in dimethylsulphoxide, which was 0·05 M in *n*-propylammonium perchlorate. (Reprinted with permission from *J. Am. chem. Soc.*, **88**, 5204, 1966.)

The usefulness of the polarographic and voltammetric half-wave potentials of reversible waves for classifying the redox behaviour of complexes and suggesting the oxidant or reductant required to effect a particular electron transfer reaction has been well illustrated by Davison and co-workers [203] using the data shown in Table 21. The compounds under study had the following formulae.

where R = C_6H_5, CF_3 or CN and M = Co, Ni, Pd, Pt or Cu.

TABLE 21

Voltammetric Data for Dithiolene Complexes in Acetonitrile Solution [195, 203]

R	Couple	$E_{\frac{1}{2}}$ (V versus the aq. S.C.E.)
[a] C_6H_5	$(Ni)^{2-} \rightleftharpoons (Ni)^- + e$	-0.74
CF_3	$(Co)^{2-} \rightleftharpoons (Co)^- + e$	-0.40
CF_3	$(Pt)^{2-} \rightleftharpoons (Pt)^- + e$	-0.27
CF_3	$(Ni)^{2-} \rightleftharpoons (Ni)^- + e$	-0.12
CN	$(Co)^{2-} \rightleftharpoons (Co)^- + e$	$+0.05$
CF_3	$(Pd)^{2-} \rightleftharpoons (Pd)^- + e$	$+0.08$
CN	$(Pt)^{2-} \rightleftharpoons (Pt)^- + e$	$+0.21$
[a] C_6H_5	$(Ni)^- \rightleftharpoons (Ni)^0 + e$	$+0.22$
CN	$(Ni)^{2-} \rightleftharpoons (Ni)^- + e$	$+0.23$
CN	$(Cu)^{2-} \rightleftharpoons (Cu)^- + e$	$+0.33$
CN	$(Pd)^{2-} \rightleftharpoons (Pd)^- + e$	$+0.44$
CF_3	$(Co)^- \rightleftharpoons (Co)^0 + e$	$+0.54$
CF_3	$(Pt)^- \rightleftharpoons (Pt)^0 + e$	$+0.82$
CF_3	$(Pd)^- \rightleftharpoons (Pd)^0 + e$	$+0.96$
CF_3	$(Ni)^- \rightleftharpoons (Ni)^0 + e$	$+1.00$

[a] Measured in dimethylsulphoxide solution.

Davison and co-workers report that all dianions in couples less positive than $\sim +0.08$ V are readily oxidized in solution (presumably by aerial oxidation), whereas all reduced species in couples more positive than this value are stable to aerial oxidation. The oxidized forms of couples more positive than $\sim +0.20$ V are unstable to reduction by solvents such as ketones and alcohols, while those in couples within the approximate range of -0.12 V to $+0.20$ V are reduced by aromatic amines such as o- and p-phenylene-diamines. Oxidized forms in couples more negative than ~ -0.12 V can be reduced by stronger reducing agents such as hydrazine or sodium amalgam; reduced forms in couples less positive than $\sim +0.44$ V can be oxidized by iodine. Finally, if an oxidant more powerful than iodine is required, bis [cis-1,2-bis(trifluoromethyl)dithiolene]nickel, $\{Ni[S_2C_2(CF_3)_2]_2\}^0$, may be employed.

A few comments on these remarks are appropriate. In strictly anhydrous solvents, all reduced species in couples more positive than -0.6 V *versus* the aq. S.C.E. should be stable to oxygen, since the formal electrode potential of the oxygen/superoxide couple in acetonitrile is ~ -0.8 V *versus* the aq. S.C.E.[40] However, in solvents containing some water, oxygen is reduced to hydrogen peroxide at a potential considerably more positive than -0.8 V; hence the reason for the above observation for aerial oxidation. The formal electrode potential of the iodine/triiodide couple in acetonitrile is about $+0.7$ V *versus* the aq. S.C.E.,[207] which fits in well with the experimental results for oxidations with iodine.

TABLE 22

Polarographic, Voltammetric and Cyclic Voltammetric Data for Reversible Waves of Miscellaneous Complexes

Redox system	Solvent and base electrolyte	Reference electrode	$E^{0\prime}$ (V *versus* R.E.)	References
$[Fe(2,2'\text{-dipy})_3]^{2+}$	CH_3CN			
$\rightarrow[complex]^+$	0.05 M Et_4NClO_4	aq. S.C.E.	-1.35	208
$\rightarrow[complex]^0$			-1.57	208
$\rightarrow[complex]^-$			-1.88	208
$[Re_2Cl_8]^{2-}$	CH_3CN			
$\rightarrow[complex]^{3-}$	0.1 M $n\text{-}Bu_4NClO_4$	aq. S.C.E.	-0.82	209
$[Re_2(NCS)_8]^{2-}$	CH_3CN			
$\rightarrow[complex]^{3-}$	0.1 M $n\text{-}Bu_4NClO_4$	aq. S.C.E.	-0.04	209
$\rightarrow[complex]^{4-}$			-0.71	209
$[Nb_6Cl_{12}]^{2+}$	DMSO			
$\rightarrow[complex]^{3+}$	0.1 M Et_4NClO_4	aq. S.C.E.	-0.10 [a]	210
$\rightarrow[complex]^{4+}$			$+0.70$ [a]	210
$\pi\text{-}(3)\text{-}1,2\text{-dicarbollyl metal}$ derivatives:				
$[(\pi\text{-}B_9C_2H_{11})_2Co]^-$	CH_3CN			
$\leftrightarrow[complex]^0$	0.1 M Et_4NClO_4	aq. S.C.E.	$+1.57$ [b]	6
$\leftrightarrow[complex]^{2-}$			-1.46 [b]	6
$[(\pi\text{-}B_9C_2H_8Br_3)_2Co]^-$	CH_3CN			
$\leftrightarrow[complex]^{2-}$	0.1 M Et_4NClO_4	aq. S.C.E.	-0.48 [b]	6
$\leftrightarrow[complex]^{3-}$			-1.58 [b]	6
$[(\pi\text{-}B_9C_2H_9Me_2)_2Co]^-$	CH_3CN			
$\leftrightarrow[complex]^{2-}$	0.1 M Et_4NClO_4	aq. S.C.E.	-1.13 [b]	6
$[(\pi\text{-}B_9C_2H_{11})_2Ni]^0$	CH_3CN			
$\leftrightarrow[complex]^-$	0.1 M Et_4NClO_4	aq. S.C.E.	$+0.25$ [b]	6
$\leftrightarrow[complex]^{2-}$			-0.59 [b]	6
$[(\pi\text{-}C_5H_5)Fe(\pi\text{-}B_9C_2H_{11})]^0$	CH_3CN			
$\leftrightarrow[complex]^-$	0.1 M Et_4NClO_4	aq. S.C.E.	-0.08 [b]	6
$[(\pi\text{-}C_5H_5)Co(\pi\text{-}B_9C_2H_{11})]^0$	CH_3CN			
$\leftrightarrow[complex]^-$	0.1 M Et_4NClO_4	aq. S.C.E.	-1.25 [b]	6
$\pi\text{-}(3)\text{-}1,7\text{-dicarbollyl metal}$ derivatives:				
$[(\pi\text{-}B_9C_2H_{11})_2Co]^-$	CH_3CN			

TABLE 22 (*contd.*)

Redox system	Solvent and base electrolyte	Reference electrode	$E^{0\prime}$ (V versus R.E.)	References
↔[complex]$^{2-}$	0·1 M Et$_4$NClO$_4$	aq. S.C.E.	−1·17 [b]	6
[(π-B$_9$C$_2$H$_{11}$)$_2$Ni]^0	CH$_3$CN			
↔[complex]$^-$	0·1 M Et$_4$NClO$_4$	aq. S.C.E.	+0·55 [b]	6
↔[complex]$^{2-}$			−0·91 [b]	6
B,B′,B″-triphenylborazine	1,2-dimethoxyethane	Ag/sat. AgNO$_3$		
↔[compound]$^-$	0·1 M Bu$_4$NClO$_4$	in glyme	−3·33	1
B,B′-diphenylborazine	1,2-dimethoxyethane,	Ag/sat. AgNO$_3$		
↔[compound]$^-$	0·1 M Bu$_4$NClO$_4$	in glyme	−3·37	1
B-phenylborazine	1,2-dimethoxyethane,	Ag/sat. AgNO$_3$		
↔[compound]$^-$	0·1 M Bu$_4$NClO$_4$	in glyme	−3·40	1
[Fe(phen)$_2$(CN)$_2$]^0	CH$_2$Cl$_2$,	Ag/Ag$^+$(?M)		
↔[complex]$^+$	0·1 M Bu$_4$NClO$_4$	0·1 M Bu$_4$NClO$_4$ in CH$_2$Cl$_2$	+0·52 [b]	2
[Fe(phen)$_2$(CNBF$_3$)$_2$]^0	CH$_2$Cl$_2$,	Ag/Ag$^+$(?M)		
↔[complex]$^+$	0·1 M Bu$_4$NClO$_4$	0·1 M Bu$_4$NClO$_4$ in CH$_2$Cl$_2$	+1·12 [b]	2
[Fe(phen)$_2$(CNBCl$_3$)$_2$]^0	CH$_2$Cl$_2$,	Ag/Ag$^+$(?M)		
↔[complex]$^+$	0·1 M Bu$_4$NClO$_4$	0·1 M Bu$_4$NClO$_4$ in CH$_2$Cl$_2$	+1·18 [b]	2
[Fe(phen)$_2$(CNBBr$_3$)$_2$]^0	CH$_2$Cl$_2$,	Ag/Ag$^+$(?M)		
↔[complex]$^+$	0·1 M Bu$_4$NClO$_4$	0·1 M Bu$_4$NClO$_4$ in CH$_2$Cl$_2$	+1·21 [b]	2

[a] No data on the reversibility of these waves are reported. They are assumed to be reversible. [b] Cyclic voltammetry was employed for these half-peak potentials; otherwise polarography or voltammetry was used. A single arrow (→) denotes that polarographic and voltammetric data were obtained only with solutions of the species in front of the arrow. A double arrow (↔) denotes that the reversibility of the system was established using cyclic voltammetry.

Organometallic Electrochemistry

Ferrocene-type organometallic compounds have already been considered in Chapter 7, Tables 8 and 9, and also in Table 19 of this chapter, but in recent extensive studies,[80, 81, 133, 211-217] Dessy and co-workers have made electrochemical investigations on many other organometallic compounds. They have employed polarography, cyclic voltammetry and controlled potential coulometry to study the electrochemical reductive behaviour of Group IV B organometallic compounds of the types Ph$_3$MX, Ph$_2$MX$_2$, Ph$_3$MM′Ph$_3$, and so on, where M or M′ is Si, Ge, Sn and Pb, and X is Cl or OAc.[133] Although most of the polarographic waves seem to be irreversible (only approximate half-wave potential values are reported), the use of these techniques has certainly led to a greater understanding of the reactions of these compounds with reducing agents. 1,2-Dimethoxyethane was used as solvent and the base electrolyte was 0·1 M tetrabutylammonium perchlorate. The original paper [133] should be consulted for full details.

Similar techniques were then employed to elucidate the reductive behaviour of organomercury compounds of the type, RHgX and R_2Hg, where R and X are alkyl or aryl groups, and halide groups respectively.[211]

These studies were continued with an investigation of the cathodic reduction of organometallic compounds of phosphorus, arsenic, antimony and bismuth. The compounds studied were Ph_3M, Ph_3MX_2, Ph_2MX, Ph_2MMPh_2, $PhMX_2$, Ph_4MX and Ph_2MOMPh_2, where M is a Group V metal and X is chloride, bromide, iodide, perchlorate or acetate. Again all waves were irreversible but by a combination of the results of polarography, cyclic voltammetry and controlled potential electrolysis with ultraviolet spectroscopy and existing chemical knowledge, the reductive behaviour of these species could be explained.[212]

Dessy and co-workers then continued their mammoth investigations by studying the electrochemical behaviour of 130 organometallic compounds of the transition elements,[80, 81] where the following approach was adopted.[80] "The normal survey of any compound involved, (1) polarographic examination, (2) multiple triangular sweep studies (that is cyclic voltammetry) to establish chemical or electrochemical reversibility in the system, (3) exhaustive controlled potential electrolysis at the appropriate potential and determination of n, the number of electrons involved in the polarographic step, (4) polarographic study of the resulting solution, (5) e.s.r. studies at this point if warranted, (6) attempted reoxidation (or reduction) of the electrochemically generated species to the initial compound, and (7) polarographic and spectroscopic studies of this final solution to compare with the initial solution." Certain of these organometallic systems were electrochemically reversible and data for these species, not recorded in earlier tables, are shown in Table 23. The reduced species were not necessarily stable for long periods in solution. Other compounds reduced irreversibly, but in certain cases the reduced species produced by controlled potential electrolysis could be oxidized by controlled potential electrolysis to the starting material with varying percentage recoveries. The original papers should be consulted for further details. The modes of reactions of organometallic compounds on electrochemical reduction are shown in Fig. 27.[80]

Electrochemical techniques were then employed by Dessy and co-workers [213] to study the scission of metal-metal bonds. The polarographic reduction waves associated with the scission of dimetallic species are irreversible, but the electrochemical reduction of homo- and hetero-dimetallic assemblies can proceed by one or other of two routes:

$$m-m' \begin{cases} \xrightarrow{2e} m:^- + m':^- \\ \xrightarrow{e} m:^- + m'\cdot \end{cases}$$

TABLE 23

Polarographic Data on Organometallic Compounds (Dessy and co-workers)

Redox system	$E^{0\prime}$ (V versus R.E.)	References
$[(2,2'\text{-dipy})Mo(CO)_4]^0 \rightarrow [complex]^-$	$-2\cdot2$	80
$[(\pi\text{-mesitylene})Mn(CO)_3]^+ \rightarrow [complex]^0$	$-0\cdot9$	80
$[(\pi\text{-mesitylene})Fe(\pi\text{-}C_5H_5)]^+ \rightarrow [complex]^0$	$-2\cdot0$	80
$[(\pi\text{-}C_6H_6)Fe(\pi\text{-}C_5H_5)]^+ \rightarrow [complex]^0$	$-1\cdot9$	80
$[(\pi\text{-}C_5H_5)Fe(CO)_2SMe]^0 \rightarrow [complex]^+$	$-0\cdot4$	80
$[(\pi\text{-}C_5H_5)Fe(CO) SMe]_2^0 \rightarrow [complex]_2^+$	$-0\cdot6$	80
$[(\pi\text{-}C_5H_5)Co S_2C_2(CF_3)_2]^0 \rightarrow [complex]^-$	$-1\cdot1$	80
$[(\pi\text{-}C_5H_5)Ni S_2C_2(CF_3)_2]^0 \rightarrow [complex]^-$	$-0\cdot96$	80
$[(\pi\text{-}C_5H_5)Rh S_2C_2(CF_3)_2]^0 \rightarrow [complex]^-$	$-1\cdot4$	80
$[(\pi\text{-anthracene})_2Cr]^+ \rightarrow [complex]^0$	$-1\cdot4$	81
$[(\pi\text{-}C_5H_5)Mo S_2C_2(CF_3)_2]_2^0 \rightarrow [complex]_2^{2-}$	$-1\cdot7$	81
$[(\pi\text{-}C_5H_5)W(CO)_2NO]^0 \rightarrow [complex]^-$	$-2\cdot4$	81
$[(\pi\text{-}C_5H_5)_2Mn_2(NO)_3]^0 \rightarrow [complex]^-$	$-1\cdot8$	81
$[(\pi\text{-cyclooctatrienone})Fe(CO)_3]^0 \rightarrow [complex]^-$	$-1\cdot7$	81
$[(\pi\text{-}C_5H_5)Co(CO)]_3^0 \rightarrow [complex]_3^-$	$-1\cdot6$	81
$[\text{Biferrocenyl}]^0 \rightarrow [complex]^+$	$0\cdot0$	81
$\rightarrow [complex]^{2+}$	$+0\cdot2$	81
$[(\pi\text{-}C_5H_5)Ir S_2C_2(CF_3)_2]^0 \rightarrow [complex]^-$	$-1\cdot7$	81
$[(\pi\text{-}C_5H_5)_3Ni_3(CO)_2]^0 \rightarrow [complex]^-$	$-1\cdot5$	81
$[(OC)_4Cr(PMe_2)]_2^0 \rightarrow [complex]_2^{2-}$	$-1\cdot8$	213
$[(OC)_3Fe(AsMe_2)]_2^0 \rightarrow [complex]_2^{2-}$	$-1\cdot9$	213

The solvent was 1,2-dimethoxyethane and the base electrolyte, 0·1 M tetrabutylammonium perchlorate; the reference electrode was silver/10^{-3} M silver perchlorate–0·1 M tetrabutylammonium perchlorate in 1,2-dimethoxyethane. This electrode has a potential of about +0·7 V versus the aq. S.C.E.

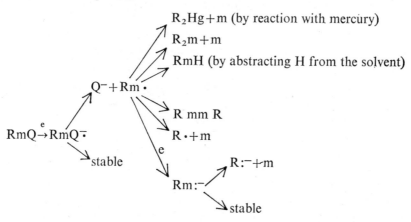

FIG. 27. The modes of reactions of organometallic compounds on electrochemical reduction. R is an alkyl, aryl moiety, etc., m is the metal, and Q is a halogen, etc.

where m and m′ are the metals and accompanying groups. In homodimetallic assemblies m = m′. Only the upper route is found for homodimetallic species. In the same paper,[213] the electrochemical reduction of polymetallics, bridged species and cluster species have been studied. A few of these species are reduced reversibly and data on them are shown in Table 23.

The investigations continued [214] with the determination of the nucleophilicities of organometallic and organometalloidal anions by comparison of their relative rates of reaction with alkyl halides. The anions were produced by electrochemical reduction of suitable compounds, usually homodimetallic compounds.[213] The decrease in concentration of the anion on reaction with the alkyl halide was determined voltammetrically using a rotating platinum electrode and a suitable potential on the plateau of the oxidation wave of the anion.

In another interesting paper, Dessy and co-workers [215] have reacted organometallic and organometalloidal anions produced by electrochemical reduction with organometallic halides or acetate. Using polarography as an analytical tool, they have established the identity of the reaction products, many of which contain metal-metal bonds. For example, $[(\pi\text{-}C_5H_5)Fe(CO)_2]^-$ reacts with Ph_3SnCl to produce $(\pi\text{-}C_5H_5)Fe(CO)_2SnPh_3$.

They have also studied redistribution reactions of the type $m:^- + m'—m'' \rightleftharpoons m—m' + m:''^-$, where m, m′ and m″ are metals and accompanying groups[216]; $m:^-$ is an organometallic anion and m′—m″ an organometallic compound containing a metal-metal bond. The organometallic anions were again produced by electrochemical reduction, and polarography was used to determine the identity of the reaction products, if any.

Finally, Psarras and Dessy [217] have carried out electrochemical oxidations and reductions of organomagnesium reagents at a dropping mercury electrode. These electron transfer processes are irreversible, but much useful information on the redox behaviour of bisorganomagnesium compounds and of Grignard reagents has been obtained. The data support a Shlenk equilibrium of $R_2Mg + MgX_2 \rightleftharpoons 2RMgX$ with $K \simeq 4$.

In conclusion, it is evident that much useful and interesting information has been obtained by Dessy and co-workers using the three relatively simple electrochemical techniques of polarography, cyclic voltammetry, and controlled potential electrolysis and coulometry. It is to be hoped that these techniques will be used more in the future by inorganic chemists than they have been in the past. A glance at the papers mentioned in this chapter will show how these techniques, combined with others such as ultraviolet, infrared, nuclear magnetic and electron spin resonance spectroscopy, can lead to a greater understanding of chemical reactions for many classes of compounds.

CHAPTER 9

Molten Salts

LEAVING aside much technical research on the production of electropositive metals such as the alkali metals and aluminium, of alloy and of metal coatings, by electrolytic reduction of molten salts,[218] most electrochemistry so far conducted for molten salts has involved fundamental investigations on transport processes, double layer effects, the kinetics of electron transfer and quantitative analysis. Considering the difficulties which exist in working with molten salts, the results of much work on electrochemistry in these solvents have been published; these results have been ably discussed in a book [219] and in articles.[218, 220-222] In a wider context, books or articles on all aspects of molten salts have been published.[223-230]

Much of the interest in inorganic electrochemistry at the moment is centred on transition metal complexes with organic ligands and on organometallic compounds. This work has been described in earlier chapters of this book. It is fairly obvious that these compounds could not be investigated in molten salts at temperatures much in excess of 300°C. In any case, it is doubtful if much would be gained by looking at the electrochemistry of these compounds in molten salts at slightly elevated temperatures rather than in aqueous or non-aqueous solvents at 25°C. However some other inorganic ions and compounds, free from organic material, might well be expected to react differently to oxidizing and reducing agents in molten salts at high temperatures than in solvents at room temperature. Certainly the electron transfer properties of these substances can be investigated in molten salts, and formal electrode potentials and n values may be determined using the techniques of direct potentiometry, polarography and voltammetry, cyclic voltammetry, chronopotentiometry and controlled potential coulometry. Of course, at temperatures in excess of 500°C, many practical problems must be overcome. Some molten salt systems which have been investigated are shown in Table 24.[231]

At high temperatures most couples behave reversibly. The dropping mercury electrode can be used up to 200°C but, at higher temperatures, stationary solid electrodes are usually employed such as carbon, platinum and tungsten electrodes.

With molten salts, an unambiguous interpretation of a voltammogram or chronopotentiogram is not always possible because there is sometimes doubt as to whether the reduced species is a pure solid or liquid, or has dissolved in the electrode and is diffusing away from the electrode surface into the electrode material. The theoretical equations for voltammograms (3.1 and 3.3) †

TABLE 24

Molten Salt Systems for Electrochemical Studies

Molten salt	Temperature (°C)
NH_4 formate	125
$LiNO_3$—$NaNO_3$—KNO_3 eut	160
$AlBr_3$ (70%)—$NaBr$ (30%)	250
LiCl—KCl eut	450
Li_2SO_4—K_2SO_4 eut	625
NaCl (50%)—KCl (50%)	700
$Na_2B_4O_7$	840
NaF—KF eut	850
NaF	1000

% = mole per cent ; eut = eutectic.

and chronopotentiograms (5.5 and 5.8) are different for each case. Another problem can arise when the species under study is reduced to a pure solid or liquid, for the activity of the metal in the first few layers of atoms to be deposited on the cathode may not be unity, and the shapes of the voltammograms and chronopotentiograms do not conform to the theoretical shapes given by Eqns (3.3) and (5.8), when the activity is other than unity. However, attempts to fit the experimental results to one model or the other are frequently successful and then the formal electrode potential is readily obtained to a good approximation. When results fit neither model, the value of the formal electrode potential cannot be determined so precisely. Of course, when both the oxidized and reduced species are present in the molten salt there can be no ambiguity and Eqns (3.1) and (5.5) must be applied.

The most widely investigated system is the lithium chloride–potassium chloride eutectic mixture at 450°C. In this solvent many formal electrode potentials have been obtained and are listed in Table 25. Electrode potentials for couples in other molten salts have been reported by Liu.[231]

† With stationary solid electrodes in molten salts, a voltammogram should exhibit a peak in the vicinity of the formal electrode potential (see p. 43). However these peaks are not found in practice and it must be assumed that steady-state conditions are rapidly achieved at stationary electrodes in molten salts as a result of convection. Hence the use of Eqns (3.1) and (3.3).

To illustrate some of the practical difficulties of working in molten salts, a voltammetric cell devised by Maricle and Hume [238] for sodium chloride (50 mole %)–potassium chloride melt at 738°C is shown in Fig. 28. With this cell, good voltammograms were obtained for silver (I), iron (II) and copper (II). The most popular reference electrodes for work in molten salts have involved the couples platinum (0)/platinum (II) and silver (0)/silver (I). However certain other reference electrodes may be used in molten salts and these have been reviewed by Laity.[239]

TABLE 25

Formal Electrode Potentials for Molten Lithium Chloride–Potassium Chloride Eutectic at 450°C

Couple	$E^{0\prime}$	References	Couple	$E^{0\prime}$	References
Li (I)—Li(0)	−3·30	232	Ni(II)—Ni(0)	−0·80	232
Mg(II)—Mg(0)	−2·58	232	V(III)—V(II)	−0·75	235
U(III)—U(0)	−2·22	233	Ag(I)—Ag(0)	−0·74	232
Mn(II)—Mn(0)	−1·85	232	HCl(g)—H$_2$(g), Pt	−0·69	237
Al(III)—Al(0)	−1·76	232	Sb(III)—Sb(0)	−0·64	232
Ti(II)—Ti(0)	−1·74	234	Bi(III)—Bi(0)	−0·55	232
Zn(II)—Zn(0)	−1·57	232	Cr(III)—Cr(II)	−0·53	232
V(II)—V(0)	−1·53	235	Hg(II)—Hg(0)	−0·5	232
Tl(I)—Tl(0)	−1·48	232	Pd(II)—Pd(0)	−0·21	232
Cr (II)—Cr(0)	−1·43	232	I(0)—I(−I)	−0·21	235
Cd(II)—Cd(0)	−1·32	232	Rh(III)—Rh(0)	−0·20	237
Ti(III)—Ti(II)	−1·32	234	UO^{2+}—UO$_2$	−0·18	236
Fe(II)—Fe(0)	−1·17	232	Ir(III)—Ir(0)	−0·06	237
U(IV)—U(III)	−1·16	236	[a] Pt(II)—Pt(0)	0·00	232
Pb(II)—Pb(0)	−1·10	232	Cu(II)—Cu(I)	+0·06	232
Sn(II)—Sn(0)	−1·08	232	Fe(III)—Fe(II)	+0·09	235
Co(II)—Co(0)	−0·99	232	Br(0)—Br(−I)	+0·18	235
Cu(I)—Cu(0)	−0·96	232	Au(I)—Au(0)	+0·21	232
Ga(III)—Ga(0)	−0·84	232	Cl(0)—Cl(−I)	+0·32	232
In(III)—In(0)	−0·80	232			

[a] The formal electrode potential of this couple was taken as 0·00.

As far as the inorganic chemist is concerned, the main interest is in knowing if new or uncommon inorganic species can be prepared and studied in molten salts. Much more work remains to be done in this field but certainly titanium (II) and uranium (III) are stable in alkali metal chloride melts (see Table 25) and zirconium (II) has been prepared by the anodic dissolution of zirconium metal in molten lithium chloride–potassium chloride eutectic at 550°C.[240] Smirnov and co-workers have also investigated the electrochemistry of beryllium (I),[241] molybdenum (II) and (III) [242, 243] and zirconium (II) [244, 245] in molten salts.

In a most interesting study in molten sodium hydroxide–potassium hydroxide eutectic mixture, in which oxide anions are the strongest base and water the strongest acid, Goret and Trémillon [246] have shown using voltammetry that oxide is oxidized first to peroxide and then to superoxide. The complete electrode potential/pH$_2$O diagram has been constructed for this system. The water-oxide product, [H$_2$O][O^{2-}], is $10^{-11.5}$ at 227°C.

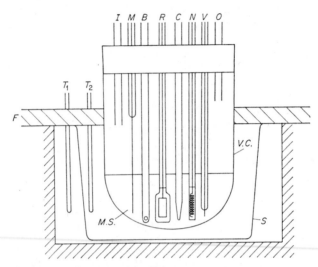

Fig 28. Cell for molten sodium chloride–potassium chloride at 738°C. F, furnace; T_1 controlling thermocouple; T_2, measuring thermocouple; S, stainless steel beaker; $V.C.$, Vycor container; $M.S.$, purified molten salt; I, inlet tube for dry argon; O, outlet tube for dry argon; B, argon bubbler made from Vycor tubing; R, reference electrode of platinum–platinum (II) in sodium chloride–potassium chloride melt. The platinum (II) was produced by electrolytic oxidation of platinum metal using a nickel electrode in nickel (II) chloride in the melt as the cathode (N). When sufficient platinum (II) had been generated, the potential of the electrode was determined against the carbon, chlorine/chloride reference electrode (C), chlorine replacing argon as the bubbled gas for this purpose. M, metal ion generating electrode sealed into borosilicate glass tubing. This is used with the nickel/nickel (II) cathode. V, the voltammetric microelectrode of tungsten sealed into Vycor tubing. The bottoms of R and N are made of Pyrex glass, which is soft at 738°C and acts as a salt bridge of resistance $\sim$50 Ω. (Reprinted with permission with slight modifications from *Analyt. Chem.*, **33**, 1188, 1961[238].)

Laitinen and Bankert [247] have indicated that solid Li$_5$CrO$_4$ can be prepared by the chronopotentiometric reduction of chromate in molten lithium chloride–potassium chloride eutectic at 450°C. As with aprotic solvents, other new oxyspecies must surely await discovery as a result of electrochemical studies in molten salts.

The stability constants of complexes can be determined in molten salts in a manner similar to that used in aqueous and non-aqueous solvents at 25°C.

For example, the stepwise stability constants of the chloro, bromo and iodo complexes of cadmium (II) have been determined in molten sodium nitrate–potassium nitrate at 263°C from chronopotentiometric measurements.[248] Stability constants for chlorocomplexes of cadmium, lead and nickel have also been determined polarographically in molten lithium nitrate–potassium nitrate eutectic at 180°C.[249]

Finally many borides, silicides, phosphides, arsenides, antimonides and sulphides have been prepared using molten salts and electrolysis.[218] For binary compounds, a salt of the metallic ion and a reducible oxyspecies of the non-metal are dissolved in the molten salt, and an anion of the non-metal is produced by electrolytic reduction of the oxyspecies. The non-metallic anions and metallic ions form an insoluble binary compound at the cathode.

Standard and Formal Electrode Potentials of Some Couples in Aqueous Solution

Couple	Medium	E^0 or $E^{0\prime}$ (V *versus* the N.H.E.)
$Li^+ + e \rightleftharpoons Li(s)$	0	$-3\cdot05$
$Cs^+ + e \rightleftharpoons Cs(s)$	0	$-2\cdot92$
$Rb^+ + e \rightleftharpoons Rb(s)$	0	$-2\cdot92$
$K^+ + e \rightleftharpoons K(s)$	0	$-2\cdot92$
$Ba^{2+} + 2e \rightleftharpoons Ba(s)$	0	$-2\cdot91$
$Sr^{2+} + 2e \rightleftharpoons Sr(s)$	0	$-2\cdot89$
$Ca^{2+} + 2e \rightleftharpoons Ca(s)$	0	$-2\cdot87$
$Na^+ + e \rightleftharpoons Na(s)$	0	$-2\cdot71$
$Ce^{3+} + 3e \rightleftharpoons Ce(s)$	0	$-2\cdot48$
$Eu^{3+} + 3e \rightleftharpoons Eu(s)$	0	$-2\cdot41$
$Mg^2 + 2e \rightleftharpoons Mg(s)$	0	$-2\cdot36$
$Th^{4+} + 4e \rightleftharpoons Th(s)$	0	$-1\cdot90$
$Be^{2+} + 2e \rightleftharpoons Be(s)$	0	$-1\cdot85$
$U^{3+} + 3e \rightleftharpoons U(s)$	0	$-1\cdot80$
$Al^{3+} + 3e \rightleftharpoons Al(s)$	0	$-1\cdot66$
$Ti^{2+} + 2e \rightleftharpoons Ti(s)$	0	$-1\cdot63$
$Cr(CN)_6^{3-} + e \rightleftharpoons Cr(CN)_6^{4-}$	0	$-1\cdot28$
$V^{2+} + 2e \rightleftharpoons V(s)$	0	$-1\cdot18$
$Mn^{2+} + 2e \rightleftharpoons Mn(s)$	0	$-1\cdot18$
$2SO_3^{2-} + 2H_2O + 2e \rightleftharpoons S_2O_4^{2-} + 4OH^-$	0	$-1\cdot12$
$Mn(CN)_6^{4-} + e \rightleftharpoons Mn(CN)_6^{5-}$	1·5 M NaCN	$-1\cdot06$
$Cr^{2+} + 2e \rightleftharpoons Cr(s)$	0	$-0\cdot91$
$Zn^{2+} + 2e \rightleftharpoons Zn(s)$	0	$-0\cdot76$
$U^{4+} + e \rightleftharpoons U^{3+}$	0	$-0\cdot61$
$As(s) + 3H^+ + 3e \rightleftharpoons AsH_3(g)$	0	$-0\cdot60$
$Ga^{3+} + 3e \rightleftharpoons Ga(s)$	0	$-0\cdot56$
$H_3PO_3 + 2H^+ + 2e \rightleftharpoons H_3PO_2 + H_2O$	0	$-0\cdot50$
$Fe^{2+} + 2e \rightleftharpoons Fe(s)$	0	$-0\cdot44$
$Eu^{3+} + e \rightleftharpoons Eu^{2+}$	1M KCl	$-0\cdot43$
$Cr^{3+} + e \rightleftharpoons Cr^{2+}$	0	$-0\cdot41$
$Cd^{2+} + 2e \rightleftharpoons Cd(s)$	0	$-0\cdot40$
$In^{3+} + 2e \rightleftharpoons In^+$	0	$-0\cdot40$

† Selected with permission from Special Publication No. 17, The Chemical Society, London.

Couple	Medium	E^0 or $E^{0\prime}$ (V *versus* the N.H.E.)
$Ti^{3+} + e \rightleftharpoons Ti^{2+}$	0	-0.37
$Se(s) + 2H^+ + 2e \rightleftharpoons H_2Se(g)$	0	-0.37
$In^{3+} + 3e \rightleftharpoons In(s)$	0	-0.34
$Tl^+ + e \rightleftharpoons Tl(s)$	0	-0.34
$Co^{2+} + e \rightleftharpoons Co(s)$	0	-0.28
$H_3PO_4 + 2H^+ + 2e \rightleftharpoons H_3PO_3 + H_2O$	0	-0.28
$V^{3+} + e \rightleftharpoons V^{2+}$	0	-0.26
$Mn(CN)_6^{3-} + e \rightleftharpoons Mn(CN)_6^{4-}$	1.5 M NaCN	-0.24
$Ni^{2+} + 2e \rightleftharpoons Ni(s)$	0	-0.23
$Sn^{2+} + 2e \rightleftharpoons Sn(s)$	0	-0.14
$Pb^{2+} + 2e \rightleftharpoons Pb(s)$	0	-0.13
$Mo^V + 2e \rightleftharpoons Mo^{III}$	0.45 M H_2SO_4	-0.01 (20°)
$H^+ + e \rightleftharpoons \frac{1}{2}H_2(g)$	0	0.00
$\frac{1}{4}P_4(s) + 3H^+ + 3e \rightleftharpoons PH_3(g)$	0	$+0.06$
$S_4O_6^{2-} + 2e \rightleftharpoons 2S_2O_3^{2-}$	0	$+0.08$
$Ti^{IV} + e \rightleftharpoons Ti^{III}$	2 M H_2SO_4	$+0.09$ (20°)
$Sn^{IV} + 2e \rightleftharpoons Sn^{II}$	2 M HCl	$+0.13$
$\frac{1}{2}Sb_2O_3(s) + 3H^+ + 3e \rightleftharpoons Sb(s) + 1\frac{1}{2}H_2O$	0	$+0.15$
$BiOCl (s) + 2H^+ + 3e \rightleftharpoons Bi(s) + H_2O + Cl^-$	0	$+0.16$
$Cu^{2+} + e \rightleftharpoons Cu^+$	0	$+0.16$
$SO_4^{2-} + 4H^+ + 2e \rightleftharpoons H_2SO_3 + H_2O$	0	$+0.17$
$S(s) + 2H^+ + 2e \rightleftharpoons H_2S(g)$	0	$+0.17$
$AgCl(s) + e \rightleftharpoons Ag(s) + Cl^-$	0	$+0.22$
$HAsO_2 + 3H^+ + 3e \rightleftharpoons As(s) + 2H_2O$	0	$+0.25$
$Hg_2Cl_2 + 2e \rightleftharpoons 2Hg(1) + 2Cl^-$	0	$+0.27$
$UO_2^{2+} + 4H^+ + 2e \rightleftharpoons U^{4+} + 2H_2O$	0	$+0.33$
$VO^{2+} + 2H^+ + e \rightleftharpoons V^{3+} + H_2O$	0	$+0.34$
$Cu^{2+} + 2e \rightleftharpoons Cu(s)$	0	$+0.34$
$Fe(CN)_6^{3-} + e \rightleftharpoons Fe(CN)_6^{4-}$	0	$+0.36$
$Mo^{VI} + e \rightleftharpoons Mo^V$	0.5 M H_2SO_4	$+0.41$ (20°)
$H_2SO_3 + 4H^+ + 4e \rightleftharpoons S(s) + 3H_2O$	0	$+0.45$
$W(CN)_8^{3-} + e \rightleftharpoons W(CN)_8^{4-}$	0	$+0.46$
$TeO_2(s) + 2H^+ + 2e \rightleftharpoons Te(s) + 2H_2O$	0	$+0.53$
$\frac{1}{2}I_2(s) + e \rightleftharpoons I^-$	0	$+0.54$
$H_3AsO_4 + 2H^+ + 2e \rightleftharpoons HAsO_2 + 2H_2O$	0	$+0.56$
$MnO_4^- + e \rightleftharpoons MnO_4^{2-}$	0	$+0.58$
$RuO_4^- + e \rightleftharpoons RuO_4^{2-}$	0	$+0.59$
$PdCl_4^{2-} + 2e \rightleftharpoons Pd(s) + 4Cl^-$	1 M HCl	$+0.62$
$O_2(g) + 2H^+ + 2e \rightleftharpoons H_2O_2$	0	$+0.68$
$Mo(CN)_8^{3-} + e \rightleftharpoons Mo(CN)_8^{4-}$	0	$+0.73$
$PtCl_6^{2-} + 2e \rightleftharpoons PtCl_4^{2-} + 2Cl^-$	0	$+0.73$
$PtCl_4^{2-} + 2e \rightleftharpoons Pt(s) + 4Cl^-$	0	$+0.73$
$H_2SeO_3 + 4H^+ + 4e \rightleftharpoons Se(s) + 3H_2O$	0	$+0.74$
$Sb^V + 2e \rightleftharpoons Sb^{III}$	3.5 M HCl	$+0.75$
$Fe^{3+} + e \rightleftharpoons Fe^{2+}$	0	$+0.77$
$Hg_2^{2+} + 2e \rightleftharpoons 2Hg(1)$	0	$+0.79$
$Ag^+ + e \rightleftharpoons Ag(s)$	0	$+0.80$

Couple	Medium	E^0 or $E^{0'}$ (V $versus$ the N.H.E.)
$NO_3^- + 2H^+ + e \rightleftharpoons \frac{1}{2}N_2O_4(g) + H_2O$	0	$+0.80$
$2Hg^{2+} + 2e \rightleftharpoons Hg_2^{2+}$	0	$+0.91$
$AuCl_4^- + 2e \rightleftharpoons AuCl_2^- + 2Cl^-$	0	$+0.93$
$NO_3^- + 4H^+ + 3e \rightleftharpoons NO(g) + 2H_2O$	0	$+0.96$
$AuCl_4^- + 3e \rightleftharpoons Au(s) + 4Cl^-$	0	$+1.00$
$RuO_4 + e \rightleftharpoons RuO_4^-$	0	$+1.00$
$VO_2^+ + 2H^+ + e \rightleftharpoons VO^{2+} + H_2O$	0	$+1.00$
$Te(OH)_6(s) + 2H^+ + 2e \rightleftharpoons TeO_2(s) + 4H_2O$	0	$+1.02$
$IrCl_6^{2-} + e \rightleftharpoons IrCl_6^{3-}$	1 M HCl	$+1.03$
$\frac{1}{2}Br_2(l) + e \rightleftharpoons Br^-$	0	$+1.07$
$SeO_4^{2-} + 4H^+ + 2e \rightleftharpoons H_2SeO_3 + H_2O$	0	$+1.15$
$ClO_4^- + 2H^+ + 2e \rightleftharpoons ClO_3^- + H_2O$	0	$+1.19$
$IO_3^- + 6H^+ + 5e \rightleftharpoons \frac{1}{2}I_2(s) + 3H_2O$	0	$+1.20$
$\frac{1}{2}O_2(g) + 2H^+ + 2e \rightleftharpoons H_2O$	0	$+1.23$
$Tl^{3+} + 2e \rightleftharpoons Tl^+$	0	$+1.26$
$PdCl_6^{2-} + 2e \rightleftharpoons PdCl_4^{2-} + 2Cl^-$	1 M HCl	$+1.29$
$\frac{1}{2}Cr_2O_7^{2-} + 7H^+ + 3e \rightleftharpoons Cr^{3+} + 3\frac{1}{2}H_2O$	0	$+1.33$
$\frac{1}{2}Cl_2(g) + e \rightleftharpoons Cl^-$	0	$+1.36$
$PbO_2(s) + 4H^+ + 2e \rightleftharpoons Pb^{2+} + 2H_2O$	0	$+1.46$
$ClO_3^- + 6H^+ + 5e \rightleftharpoons \frac{1}{2}Cl_2(g) + 3H_2O$	0	$+1.47$
$MnO_4^- + 8H^+ + 5e \rightleftharpoons Mn^{2+} + 4H_2O$	0	$+1.51$
$BrO_3^- + 6H^+ + 5e \rightleftharpoons \frac{1}{2}Br_2(l) + 3H_2O$	0	$+1.52$
$Ce^{4+} + e \rightleftharpoons Ce^{3+}$	1 M HClO$_4$	$+1.74$
$H_2O_2 + 2H^+ + 2e \rightleftharpoons 2H_2O$	0	$+1.77$
$Co^{3+} + e \rightleftharpoons Co^{2+}$	3 M HNO$_3$	$+1.84$
$Ag^{2+} + e \rightleftharpoons Ag^+$	4 M HClO$_4$	$+2.00$
$S_2O_8^{2-} + 2e \rightleftharpoons 2SO_4^{2-}$	0	$+2.01$
$O_3(g) + 2H^+ + 2e \rightleftharpoons O_2(g) + H_2O$	0	$+2.07$
$\frac{1}{2}F_2(g) + e \rightleftharpoons F^-$	0	$+2.87$

The temperature is 25° C unless otherwise stated. Where "0" appears in the second column, the value in the third column is the standard electrode potential; in all other cases the value in the third column is the formal electrode potential.

References

1. Shriver, D. F., Smith, D. E. and Smith, P. (1964). *J. Am. chem. Soc.* **86,** 5153.
2. Shriver, D. F. and Posner, J. (1966). *J. Am. chem. Soc.* **88,** 1672.
3. Olsen, D. C., Mayweg, V. P. and Schrauzer, G. N. (1966). *J. Am. chem. Soc.* **88,** 4876.
4. Hall, D. W. and Russell, C. D. (1967). *J. Am. chem. Soc.* **89,** 2316.
5. Coetzee, J. F., McGuire, D. K. and Hedrick, J. L. (1963). *J. Phys. Chem.* **67,** 1814.
6. Hawthorne, M. F., Young, D. C., Andrews, T. D., Howe, D. V., Pilling, R. L., Pitts, A. D., Reintjes, M., Warren, L. F. jr. and Wegner, P. A. (1968). *J. Am. chem. Soc.* **90,** 879.
7. Latimer, W. M. (1952). "Oxidation Potentials" (second edition), p. 99. Prentice-Hall, New Jersey.
8. Swift, E. H. (1939). "A System of Chemical Analysis." Prentice-Hall, New Jersey.
9. Bock, R. and Herrmann, M. (1953). *Z. anorg. all. Chem.* **273,** 1.
10. *Stability Constants of Metal-Ion Complexes* (1964). Special Publication, No. 17. The Chemical Society, London.
11. Perrin, D. D. (1959). *Rev. pure appl. Chem.* **9,** 257.
12. Vlček, A. A. (1963). *Polarographic behaviour of coordination compounds.* In "Progress in Inorganic Chemistry" (ed. Cotton), Vol. 5, p. 211. Interscience, New York.
13. Vlček, A. A. (1962). *Mechanism of the electrode processes and structure of inorganic complexes.* In "Progress in Polarography" (ed. Zuman), Vol. 1, p. 269. Interscience, New York.
14. Delahay, P. (1954). *Discuss. Faraday Soc.* **17,** 205.
15. Nicholson, R. S. and Shain, I. (1964). *Analyt. Chem.* **36,** 706.
16. Meites, L. (1961). *Rec. chem. Prog.* **22,** 81.
17. Kolthoff, I. M. and Lingane, J. J. (1952). "Polarography" (Vols 1 and 2). Interscience, New York.
18. Milner, G. W. C. (1957). "The Principles and Application of Polarography and other Electroanalytical Processes." Longmans, London.
19. Meites, L. (1965). "Polarographic Techniques" (second edition). Interscience, New York.
20. Crow, D. R. and Westwood, J. V. (1968). "Polarography." Methuen, London.
21. Lingane, J. J. (1958). "Electroanalytical Chemistry" (second edition). Interscience, New York.
22. Rechnitz, G. A. (1963). "Controlled-Potential Analysis." Pergamon Press, Oxford.
23. "Treatise on Analytical Chemistry" (1963) (eds. I. M. Kolthoff and P. J. Elving), Part 1, Vol. **4,** D-2, p. 2109. Wiley, New York.
24. Wenglowski, G. M. (1966). *An economic study of the electrochemical industry in the United States.* In "Modern Aspects of Electrochemistry" (ed. Bockris), Vol. 4, p. 251. Butterworths, London.

25. Rossotti, F. J. C. and Rossotti, H. (1961). "The Determination of Stability Constants", Chapter 7. McGraw-Hill, New York.
26. Meites, L. (1965). "Polarographic Techniques" (second edition), p. 88. Interscience, New York.
27. Taylor, M. S. (1963). Ph.D. Thesis, p. 82. University of Sheffield.
28. Lingane, J. J. (1958). "Electroanalytical Chemistry" (second edition), p. 209. Interscience, New York.
29. Kumar, G. P. and Pantony, D. A. (1966). "Polarography 1964" (ed. Hills), p. 1061. Macmillan, London.
30. Headridge, J. B. and Pletcher, D. (1966). *J. chem. Soc.* 757.
31. Callingham, M. and Headridge, J. B. Unpublished results.
32. Ashraf, M., Headridge, J. B. and Pletcher, D. Unpublished results.
33. Hills, G. J. (1961). *Reference electrodes in nonaqueous solutions.* In "Reference Electrodes" (eds Ives and Janz), p. 433. Academic Press, New York.
34. Takahashi, R. (1965). *Talanta*, **12**, 1211.
35. Latimer, W. M. (1952). "Oxidential Potentials" (second edition), p. 7. Prentice-Hall, New Jersey.
36. Popov, A. I. (1963). *Techniques with nonaqueous solvents.* In "Technique of Inorganic Chemistry" (eds. Jonassen and Weissberger), Vol. 1, p. 82. Interscience, New York.
37. Mason, J. G. and Rosenblum, M. (1960). *J. Am. chem. Soc.* **82**, 4206.
38. Davis, H. M. and Rooney, R. C. (1962). *J. polarogr. Soc.* **8**, 25.
39. Barker, G. C. and Gardner, A. W. (1960). *Z. analyt. Chem.* **173**, 79.
40. Peover, M. E. and White, B. S. (1965). *Chem. Comm.* 183.
41. Johnson, E. J., Pool, K. H. and Hamm, R. E. (1966). *Analyt. Chem.* **38**, 183.
42. Buck, R. P. (1963). *J. electroanal. Chem.* **5**, 295.
43. Meites, L. (1965). "Polarographic Techniques" (second edition), p. 177. Interscience, New York.
44. Katz, T. J., Reinmuth, W. H. and Smith, D. E. (1962). *J. Am. chem. Soc.* **84**, 802.
45. Meites, L. (1965). "Polarographic Techniques" (second edition), pp. 227 and 253. Interscience, New York.
46. Crow, D. R. and Westwood, J. V. (1965). *Q. Rev. chem. Soc.* **19**, 57.
47. Hume, D. N., DeFord, D. D. and Cave, G. C. B. (1951). *J. Am. chem. Soc.* **73**, 5323.
48. Reilley, C. N. (1963). *Fundamentals of electrode processes.* In "Treatise on Analytical Chemistry" (eds. Kolthoff and Elving), Part 1, Vol. 4, Chap. 42, p. 2109. Interscience, New York.
49. Jordan, J. and Stalica, N. R. (1963). *Kinetics of heterogeneous electron transfer in electrode processes.* In "Handbook of Analytical Chemistry" (ed. Meites), pp. 5–38. McGraw-Hill, New York.
50. Bauer, H. H. (1968). *J. electroanalyt. Chem.* **16**, 419.
51. Meites, J. (1965). "Polarographic Techniques" (second edition), p. 236. Interscience, New York.
52. Damaskin, B. B. (1967). "The Principles of Current Methods for the Study of Electrochemical Reactions." McGraw-Hill, New York.
53. Reynolds, W. L. and Lumry, R. W. (1966). "Mechanisms of Electron Transfer." Ronald Press, New York.
54. Sykes, A. G. (1966). "Kinetics of Inorganic Reactions," Chaps. 7 and 8. Pergamon Press. Oxford.

55. Basolo, F. and Pearson, R. G. (1967). "Mechanisms of Inorganic Reactions" (second edition), Chap. 6. Wiley, New York.
56. Edwards, J. O. (1964). "Inorganic Reaction Mechanisms," Chap. 7. Benjamin, New York.
57. Sykes, A. G. (1967). *Further advances in the study of mechanisms of redox reactions.* In "Adv. in Inorg. Chem. and Radiochem." (eds. Eméleus and Sharpe), Vol. 10, p. 153. Academic Press, New York.
58. Dickens, J. E., Basolo, F. and Neumann, H. M. (1957). *J. Am. chem. Soc.* **79,** 1286.
59. Basolo, F., Hayes, J. C. and Neumann, H. M. (1953). *J. Am. chem. Soc.* **75,** 5102.
60. Margerum, D. W. (1957). *J. Am. chem. Soc.* **79,** 2728.
61. Baxendale, J. H. and George, P. (1950). *Trans. Faraday Soc.* **46,** 736.
62. Emschwiller, G. and Le Gros, J. (1955). *C.r. hebd. Séanc. Acad. Sci. Paris* **241,** 44.
63. Wilkins, R. G. and Williams, M. J. G. (1957). *J. chem. Soc.* 4514.
64. Stranks, D. R. (1960). *Discuss. Faraday Soc.* **29,** 73.
65. Dietrich, M. W. and Wahl, A. C. (1963). *J. chem. Phys.* **38,** 1591.
66. Kruse, W. and Taube, H. (1960). *J. Am. chem. Soc.* **82,** 526.
67. Sullivan, J. C., Cohen, D. and Hindman, J. C. (1957). *J. Am. chem. Soc.* **79,** 3672.
68. Wahl, A. C. (1960). *Z. Elektrochem.* **64,** 90.
69. Lewis, W. B., Coryell, C. D. and Irving, J. W. (1949). *J. chem. Soc.* S.386.
70. Taube, H., Myers, H. and Rich, R. C. (1953). *J. Am. chem. Soc.* **75,** 4118.
71. Ogard, A. E. and Taube, H. (1958). *J. Am. chem. Soc.* **80,** 1084.
72. Marcus, R. A. (1963). *J. phys. Chem.* **67,** 853.
73. Marcus, R. A. (1964). *A. Rev. phys. Chem.* 155.
74. Ross, J. W., DeMars, R. D. and Shain, I. (1956). *Analyt. Chem.* **28,** 1768.
75. Frankenthal, R. P. and Shain, I. (1956). *J. Am. chem. Soc.* **78,** 2969.
76. Randles, J. E. B. (1948). *Trans. Faraday Soc.* **44,** 327.
77. Ševčík, A. (1948). *Coll. Czech. chem. Comm.* **13,** 349.
78. Nicholson, M. M. (1954). *J. Am. chem. Soc.* **76,** 2539.
79. Nicholson, R. S. (1965). *Analyt. Chem.* **37,** 1351.
80. Dessy, R. E., Stary, F. E., King, R. B. and Waldrop, M. (1966). *J. Am. chem. Soc.* **88,** 471.
81. Dessy, R. E., King, R. B. and Waldrop, M. (1966). *J. Am. chem. Soc.* **88,** 5112.
82. Rupp, E. B., Smith, D. E. and Shriver, D. F. (1967). *J. Am. chem. Soc.* **89,** 5562.
83. Smith, D. E., Rupp, E. B. and Shriver D. F. (1967). *J. Am. chem. Soc.* **89,** 5568.
84. Middaugh, R. L. and Farha, F. jr. (1966). *J. Am. chem. Soc.* **88,** 4147.
85. Kuwana, T., Bublitz, D. E. and Hoh, G. L. K. (1960). *J. Am. chem. Soc.* **82,** 5811.
86. McClure, J. E. and Maricle, D. L. (1967). *Analyt. Chem.* **39,** 236.
87. Delahay, P. (1954). "New Instrumental Methods in Electrochemistry", Chap. 8. Interscience, New York.
88. Lingane, J. J. (1958). "Electroanalytical Chemistry" (second edition), chap. 22, Interscience, New York.
89. Meites, L. (1965). "Polarographic Techniques" (second edition), chap. 10, section 7. Interscience, New York.

90. Paunovic, M. (1967). *J. electroanal. Chem.* **14,** 447.
91. Meites, L. (1965). "Polarographic Techniques" (second edition), pp. 582, 586–588. Interscience, New York.
92. Kaldova, R. (1962). *Oscillographic polarography with alternating current.* In "Progress in Polarography" (ed. Zuman), Vol. 2, p. 449. Interscience, New York.
93. Aten, A. C., Biithker, C. and Hoijtink, G. J. (1959). *Trans. Faraday Soc.* **55,** 324.
94. Breyer, B. and Bauer, H. H. (1963). "Alternating Current Polarography and Tensammetry". Interscience, New York.
95. Heyrovský, J. and Kůta, J. (1966). "Principles of Polarography," Chap. 21. Academic Press, New York.
96. Meites, L. (1965). "Polarographic Techniques" (second edition), p. 511. Interscience, New York.
97. Lott, P. F. (1965). *J. chem. Educ.* **42,** A261 and A361.
98. Rechnitz, G. A. (1962). *Inorg. Chem.* **1,** 953.
99. Balch, A. L. and Holm, R. H. (1966). *J. Am. chem. Soc.* **88,** 5201.
100. Holm, R. H., Balch, A. L., Davison, A., Maki, A. H. and Berry, T. E. (1967). *J. Am. chem. Soc.* **89,** 2866.
101. McCleverty, J. A., Atherton, N. M., Locke, J., Wharton, E. J. and Winscom, C. J. (1967). *J. Am. chem. Soc.* **89,** 6082.
102. Kies, H. L. (1962). *J. electroanal. Chem.* **4,** 257.
103. Latimer, W. M. (1952). "Oxidation Potentials" (second edition), Prentice-Hall, New Jersey.
104. Parsons, R. (1959). "Handbook of Electrochemical Constants." Butterworths, London.
105. de Bethune, A. J. and Loud, N. A. S. (1964). "Standard Aqueous Electrode Potentials and Temperature Coefficients at 25°C." Hampel, Stokie, Illinois.
106. Meites, L. (1963). *Polarographic characteristics of inorganic substances.* In "Handbook of Analytical Chemistry" (ed. Meites), pp. 5–50. McGraw-Hill, New York.
107. Page, J. A. and Wilkinson, G. (1952). *J. Am. chem. Soc.* **74,** 6149.
108. Cotton, F. A., Whipple, R. O. and Wilkinson, G. (1953). *J. Am. chem. Soc.* **75,** 3586.
109. Wilkinson, G., Pauson, P. L. and Cotton, F. A. (1954). *J. Am. chem. Soc.* **76,** 1970.
110. Wilkinson, G., Pauson, P. L., Birmingham, J. M. and Cotton, F. A. (1953). *J. Am. chem. Soc.* **75,** 1011.
111. Wilkinson, G. and Birmingham, J. M. (1954). *J. Am. chem. Soc.* **76,** 4281.
112. Pauson, P. L. and Wilkinson, G. (1954). *J. Am. chem. Soc.* **76,** 2024.
113. Korshunov, I. A., Vertyulina, L. N., Razuvaev, G. A., Sorokin, Y. A. and Domrachev, G. A. (1958). *Dokl. Akad. Nauk SSSR,* **122,** 1029.
114. Koepp, Von H.-M., Wendt, H. and Strehlow, H. (1960). *Z. Elektrochem.* **64,** 483.
115. Smith, G. F. and Richter, F. P. (1944). "Phenanthroline and Substituted Phenanthroline Indicators." G. F. Smith, Columbus, Ohio.
116. Smith, G. F. and Brandt, W. W. (1949). *Analyt. Chem.* **21,** 948.
117. Nelson, I. V. and Iwamoto, R. T. (1961). *Analyt. Chem.* **33,** 1795.
118. Nelson, I. V. and Iwamoto, R. T. (1963). *Analyt. Chem.* **35,** 867.
119. Brandt, W. W. and Smith, G. F. (1949). *Analyt. Chem.* **21,** 1313.

120. Korshunov, I. A., Vertyulina, L. N. and Domrachev, G. A. (1962). *Zh. Obshch. Khim.* **32**, 10.
121. Brill, A. S., Martin, R. B. and Williams, R. J. P. (1964). In "Electronic Aspects of Biochemistry" (ed. Pullman), p. 519. Academic Press, New York.
122. James, B. R., Lyons, J. R. and Williams, R. J. P. (1962). *Biochem.* **1**, 379.
123. Tirouflet, J., Laviron, E., Dabard, R. and Komenda, J. (1963). *Bull. Soc. chim. Fr.*, 857.
124. Vlček, A. A. (1965). *Coll. Czech. chem. Comm.* **30**, 952.
125. Tomkinson, J. C. and Williams, R. J. P. (1958). *J. Chem. Soc.* 2010.
126. Komenda, J. and Tirouflet, J. (1962). *C.r. hebd. Séanc. Acad. Sci. Paris*, **254**, 3093.
127. Gubin, S. P. and Perevalova, E. G. (1962). *Dokl. Akad. Nauk SSSR*, **143**, 1351.
128. Kolthoff, I. M. and Reddy, T. B. (1961). *J. electrochem. Soc.* **108**, 980.
129. Headridge, J. B., Ashraf, M. and Dodds, H. L. H. (1968). *J. electroanal. Chem.* **16**, 114.
130. Headridge, J. B. and Pletcher, D. (1967). *J. electroanal. Chem.* **15**, 312.
131. Kolthoff, I. M. and Coetzee, J. F. (1957). *J. Am. chem. Soc.* **79**, 870.
132. Kolthoff, I. M. and Coetzee, J. F. (1957). *J. Am. chem. Soc.* **79**, 1852.
133. Dessy, R. E., Kitching, W. and Chivers, T. (1966). *J. Am. chem. Soc.* **88**, 453.
134. Ashraf, M. and Headridge, J. B. Unpublished results.
135. Headridge, J. B., Pletcher, D. and Callingham, M. (1967). *J. chem. Soc.* 684.
136. Coetzee, J. F. and Siao, W.-S. (1963). *Inorg. Chem.* **2**, 14.
137. Voorhies, J. D. and Schurdak, E. J. (1962). *Analyt. Chem.* **34**, 939.
138. McCleverty, J. A. and Wharton, E. J. Unpublished results.
139. Cisak, A. and Elving, P. J. (1963). *J. electrochem. Soc.* **110**, 160.
140. Turner, W. R. and Elving, P. J. (1965). *Analyt. Chem.* **37**, 467.
141. Headridge, J. B. Personal observation.
142. Kratochvil, B., Zatko, D. A. and Markuszewski, R. (1966). *Analyt. Chem.* **38**, 770.
143. Farha, F. jr. and Iwamoto, R. T. (1964). *J. electroanal. Chem.* **8**, 55.
144. Hoijtink, G. J., De Boer, E., Van der Meij, P. H. and Weijland, W. P. (1956). *Recl. Trav. chim. Pays-Bas Belg.* **75**, 487.
145. Shrivington, P. J. (1967). *Aust. J. Chem.* **20**, 447.
146. Forcier, G. A. and Olver, J. W. (1965). *Analyt. Chem.*, **37**, 1447.
147. Moe, N. S. (1967). *Acta Chem. Scand.* **21**, 1389.
148. Peover, M. E. (1964). *Trans. Faraday Soc.* **60**, 417.
149. Charlot, G., Badoz-Lambling, J. and Trémillon, B. (1962). "Electrochemical Reactions". Elsevier, Amsterdam.
150. Popov, A. I. (1963). *Techniques with nonaqueous solvents.* In "Technique of Inorganic Chemistry" (ed. Jonassen and Weissberger), Vol. 1, p. 37. Interscience, New York.
151. Butler, J. N. (1967). *J. electroanal. Chem.* **14**, 89.
152. Meites, L. (1963). "Handbook of Analytical Chemistry" (ed. Meites), pp. 5–15. McGraw-Hill, New York.
153. Vlček, A. A. (1951). *Coll. Czech. chem. Comm.* **16**, 230.
154. Strehlow, H. (1952). *Z. Elektrochem.* **56**, 827.
155. Coetzee, J. F. and Campion, J. J. (1967). *J. Am. chem. Soc.* **89**, 2513.
156. Coulter, R. D. T. and Iwamoto, R. T. (1967). *J. electroanal. Chem.* **13**, 21.
157. Knecht, L. A. and Kolthoff, I. M. (1962). *Inorg. Chem.* **1**, 195.
158. McMasters, D. L., Dunlap, R. B., Kuempel, J. R., Kreider, L. W. and Shearer, T. R. (1967). *Analyt. Chem.* **39**, 103.

159. Číhalík, J. and Šimek, J. (1958). *Coll. Czech. chem. Comm.* **23**, 615.
160. Meites, L. (1955). "Polarographic Techniques" (first edition). Interscience, New York.
160a. Headridge, J. B., Hamza, A. G., Hubbard, D. P. and Taylor, M. S. (1966). "Polarography 1964" (ed. Hills), p. 625. Macmillan, London.
161. Headridge, J. B. and Pletcher, D. Unpublished results.
162. Drago, R. S. and Purcell, K. F. (1964). *The coordination model for non-aqueous solvent behaviour*. In "Progress in Inorganic Chemistry" (ed. Cotton), Vol. 6, p. 294. Interscience, New York.
163. Gutmann, V., Steininger, A. and Wychera, E. (1966). *Mh. Chem.* **97**, 460.
164. Gutmann, V. and Mayer, U. (1967). *Mh. Chem.* **98**, 294.
165. Kolthoff, I. M. and Thomas, F. G. (1965). *J. phys. Chem.* **69**, 3049.
166. Kolthoff, I. M. and Lingane, J. J. (1952). "Polarography", Vol. 1, p. 227. Interscience, New York.
167. Biedermann, G. and Wallin, T. (1960). *Acta. chem. Scand.* **14**, 594.
168. Cozzi, D. and Vivarelli, S. (1954). *Z. Elektrochem.* **58**, 907.
169. Auerbach, C. and McGuire, D. K. (1966). *J. inorg. nucl. Chem.* **28**, 2659.
170. Ciana, A. and Furlani, C. (1965). *Electrochim. Acta*, **10**, 1149.
171. Larson, R. C. and Iwamoto, R. T. (1962). *Inorg. Chem.* **1**, 316.
172. Nelson, I. V. and Iwamoto, R. T. (1964). *Inorg. Chem.* **3**, 661.
173. Elving, P. J. and Spritzer, M. S. (1965). *Talanta*, **12**, 1243.
174. Spritzer, M. S., Costa, J. M. and Elving, P. J. (1965). *Analyt. Chem.* **37**, 211.
175. Kolthoff, I. M. (1964). *J. polarogr. Soc.* **10**, 22.
176. Peover, M. E. and White, B. S. (1966). *Electrochim. Acta*, **11**, 1061.
177. Sawyer, D. T. and Roberts, J. L. (1966). *J. electroanal. Chem.* **12**, 90.
178. Maricle, D. L. and Hodgson, W. G. (1965). *Analyt. Chem.* **37**, 1562.
179. Goolsby, A. D. and Sawyer, D. T. (1968). *Analyt. Chem.* **40**, 83.
180. Meites, L. (1955). "Polarographic Techniques" (first edition), p. 279. Interscience, New York.
181. Michlmayr, M., Gritzner, G. and Gutmann, V. (1966). *Inorg. nucl. Chem. Letters*, **2**, 227.
182. Headridge, J. B. and Pletcher, D. (1967). *J. polarogr. Soc.* **13**, 107.
183. Furlani, C. and Fischer, E. O. (1957). *Z. Elektrochem.* **61**, 481.
184. Furlani, C. and Sartori, G. (1958). *Ricerca scient.* **28**, 973.
185. Furlani, C. (1966). *Ricerca scient.* **36**, 989.
186. Furlani, C., Furlani, A. and Sestili, I. (1965). *J. electroanal. Chem.* **9**, 140.
187. Bublitz, D. E., Hoh, G. and Kuwana, T. (1959). *Chemy. Ind.* 635.
188. Hsiung, H. and Brown, G. H. (1963). *J. electrochem. Soc.* **110**, 1085.
189. Valcher, S. and Mastragostino, M. (1967). *J. electroanal. Chem.* **14**, 219.
190. Gulick, W. M. jr. and Geske, D. H. (1967). *Inorg. Chem.* **6**, 1320.
191. Schroer, H. P. and Vlček, A. A. (1964). *Z. anorg. allgem. Chem.* **334**, 205.
192. Hoh, G. L. K., McEwen, W. E. and Kleinberg, J. (1961). *J. Am. chem. Soc.* **83**, 3949.
193. Little, W. F., Reilley, C. N., Johnson, J. D., Lynn, K. N. and Sanders, A. P. (1964). *J. Am. chem. Soc.* **86**, 1376.
194. Little, W. F., Reilley, C. N., Johnson, J. D. and Sanders, A. P. (1964). *J. Am. chem. Soc.* **86**, 1382.
195. Davison, A., Edelstein, N., Holm, R. H. and Maki, A. H. (1963). *Inorg. Chem.* **2**, 1227.
196. Davison, A., Edelstein, N., Holm, R. H. and Maki, A. H. (1964). *J. Am. chem. Soc.* **86**, 2799.

197. Davison, A., Edelstein, N., Holm, R. H. and Maki, A. H. (1965). *Inorg. Chem.* **4,** 55.
198. Balch, A. L., Rohrscheid, F. and Holm, R. H. (1965). *J. Am. chem. Soc.* **87,** 2301.
199. Williams, R., Billig, E., Waters, J. H. and Gray, H. B. (1966). *J. Am. chem. Soc.* **88,** 43.
200. Baker-Hawkes, M. J., Billig, E. and Gray, H. B. (1966). *J. Am. chem. Soc.* **88,** 4870.
201. Stiefel, E. J. and Gray, H. B. (1965). *J. Am. chem. Soc.* **87,** 4012.
202. Balch, A. L. and Holm, R. H. (1966). *Chem. Comm.* 552.
203. Davison, A., Edelstein, N., Holm, R. H. and Maki, A. H. (1964). *Inorg. Chem.* **3,** 814.
204. Schrauzer, G. N., Mayweg, V. P., Finck, H. W. and Heinrich, W. (1966). *J. Am. chem. Soc.* **88,** 4604.
205. Davison, A., McCleverty, J. A., Shawl, E. T. and Wharton, E. J. (1967). *J. Am. chem. Soc.* **89,** 830.
206. McCleverty, J. A. (1968). *Metal 1,2-dithiolene and related complexes.* In "Progress in Inorganic Chemistry" (ed. Cotton), Vol. 10, p. 49. Interscience, New York.
207. Popov, A. I. and Geske, D. H. (1958). *J. Am. chem. Soc.* **80,** 1340.
208. Tanaka, N. and Sato, Y. (1966). *Inorg. nucl. Chem. Lett.* **2,** 359.
209. Cotton, F. A., Robinson, W. R. and Walton, R. A. (1967). *Inorg. Chem.* **6,** 1257.
210. Mackay, R. A. and Schneider, R. F. (1967). *Inorg. Chem.* **6,** 549.
211. Dessy, R. E., Kitching, W., Psarras, T., Salinger, R., Chen, A. and Chivers, T. (1966). *J. Am. chem. Soc.* **88,** 460.
212. Dessy, R. E., Chivers, T. and Kitching, W. (1966). *J. Am. chem. Soc.* **88,** 467.
213. Dessy, R. E., Weissman, P. M. and Pohl, R. L. (1966). *J. Am. chem. Soc.* **88,** 5117.
214. Dessy, R. E., Pohl, R. L. and King, R. B. (1966). *J. Am. chem. Soc.* **88,** 5121.
215. Dessy, R. E. and Weissman, P. M. (1966). *J. Am. chem. Soc.* **88,** 5124.
216. Dessy, R. E. and Weissman, P. M. (1966). *J. Am. chem. Soc.* **88,** 5129.
217. Psarras, T. and Dessy, R. E. (1966). *J. Am. Chem. Soc.* **88,** 5132.
218. Reddy, T. B. (1963). *Electro. Technol.* **1,** 325.
219. Delimarskii, Iu. K. and Markov, B. F. (1961). "Electrochemistry of Fused Salts." Sigma Press, Washington.
220. Laitinen, H. A. and Osteryoung, R. A. (1964). *Electrochemistry in molten salts.* In "Fused Salts" (ed. Sundheim), p. 255. McGraw-Hill, New York.
221. Liu, C. H., Johnson, K. E. and Laitinen, H. A. (1964). *Electroanalytical chemistry of molten salts.* In "Molten Salt Chemistry" (ed. Blander), p. 681. Interscience, New York.
222. Graves, A. D., Hills, G. J. and Inman, D. (1966). *Electrode processes in molten salts.* In "Adv. in Electrochem. and Electrochem. Eng." (ed. Delahay), Vol. 4, p. 117. Interscience, New York.
223. Bloom, H. and Bockris, J. O'M. (1959). *Molten Electrolytes.* In "Modern Aspects of Electrochemistry" (ed. Bockris), No. 2, p. 160. Butterworths, London.
224. Corbett, J. D. and Duke, F. R. (1963). *Fused salt techniques.* In "Technique of Inorganic Chemistry" (eds Jonassen and Weissberger), Vol. 1, p. 103. Interscience, New York.

225. Blander, M., editor (1964). "Molten Salt Chemistry." Interscience, New York.
226. Corbett, J. D. (1964). *Fused salt chemistry*. In "Survey of Progress in Chemistry" (ed. Scott), Vol. 2, p. 91. Academic Press. New York.
227. Sundheim, B. S., editor (1964). "Fused Salts". McGraw-Hill, New York.
228. Bloom, H. and Hastie, J. W. (1965). *Molten salts as solvents*. In "Non-aqueous Solvent Systems" (ed. Waddington), p. 353. Academic Press, New York.
229. Bailey, R. A. and Janz, G. J. (1966). *Experimental techniques in the study of fused salts*. In "The Chemistry of Non-Aqueous Solvents" (ed. Lagowski), Vol. 1, p. 292. Academic Press, New York.
230. Bloom, H. (1968). "The Chemistry of Molten Salts." Benjamin, New York.
231. Liu, C. H. (1963). *Electroanalytical chemistry in fused-salt media*. In "Handbook of Analytical Chemistry" (ed. Meites), pp. 5–218. McGraw-Hill, New York.
232. Laitinen, H. A. and Liu, C. H. (1958). *J. Am. chem. Soc.* **80**, 1015.
233. Gruen, D. M. and Osteryoung, R. A. (1960). *Ann. N.Y. Acad. Sci.* **79**, 897.
234. Baboian, R., Hill, D. L. and Bailey, R. A. (1965). *Can. J. Chem.* **43**, 197.
235. Laitinen, H. A. and Pankey, J. W. (1959). *J. Am. chem. Soc.*, **81**, 1053.
236. Hill, D. L., Perano, J. and Osteryoung, R. A. (1960). *J. electrochem. Soc.* **107**, 698.
237. Laitinen, H. A. and Plambeck, J. A. (1965). *J. Am. chem. Soc.* **87**, 1202.
238. Maricle, D. L. and Hume, D. N. (1961). *Analyt. Chem.* **33**, 1188.
239. Laity, R. W. (1961). *Electrodes in fused salt systems*. In "Reference Electrodes" (eds Ives and Janz), p. 524. Academic Press, New York.
240. Baboian, R. (1965). *Diss. Abstr.* **26**, 1362.
241. Smirnov, M. V. and Chukreev, N. Ya. (1962). *Tr. Inst. Elektrokhim.* No. 3, 3. Akad. Nauk SSSR, Ural'sk Filial,
242. Smirnov, M. V. and Ryzhik, O. A. (1966). "Electrochemistry of Molten and Solid Electrolytes" (ed. Baraboshkin), Vol. 3, p. 9. Consultants Bureau, New York.
243. Ryzhik, O. A. and Smirnov, M. V. (1967). "Electrochemistry of Molten and Solid Electrolytes" (eds. Baraboshkin and Pal'guev), Vol. 4, pp. 21 and 27. Consultants Bureau, New York.
244. Komarov, V. E., Smirnov, M. V. and Baraboshkin, A. N. (1961). "Electrochemistry of Molten and Solid Electrolytes," Vol. 1, p. 16. Consultants Bureau, New York.
245. Kudyakov, V. Ya. and Smirnov, M. V. (1966). "Electrochemistry of Molten and Solid Electrolytes" (ed. Baraboshkin), Vol. 3, p. 15. Consultants Bureau, New York.
246. Goret, J. and Trémillon, B. (1966). *Bull. Soc. chim. Fr.* 67.
247. Laitinen, H. A. and Bankert, R. D. (1967). *Analyt. Chem.* **39**, 1790.
248. Inman, D. and Bockris, J. O'M. (1961). *Trans. Faraday Soc.* **57**, 2308.
249. Christie, J. H. and Osteryoung, R. A. (1960). *J. Am. chem. Soc.* **82**, 1841.
250. Buckingham, D. A. and Sargeson, A. M. (1964). *Oxidation-reduction potentials as functions of donor atom and ligand*. In "Chelating Agents and Metal Chelates" (eds Dwyer and Mellor), p. 237. Academic Press, New York.
251. Falk, J. E. and Phillips, J. N. (1964). *Redox behaviour of metalloporphyrins*. In "Chelating Agents and Metal Chelates" (eds Dwyer and Mellor), p. 468. Academic Press, New York.

Author Index

Numbers in parentheses are reference numbers and are included to assist in locating references. Full references are cited in numerical order on pp. 109–116.

Subject Index